Adios Amigos

Adios Amigos

*Tales of Sustenance and Purification
in the American West*

Page Stegner

COUNTERPOINT · BERKELEY

Copyright © 2008 by Page Stegner. All rights reserved under
International and Pan-American Copyright Conventions.

Library of Congress Cataloging-in-Publication Data

Stegner, Page.
Adios amigos : tales of sustenance and purification in the
American West / Page
Stegner.
p. cm.
ISBN-13: 978-1-59376-169-1
ISBN-10: 1-59376-169-4
1. West (U.S.)—Description and travel—Anecdotes. 2.
Wild and scenic rivers—West (U.S.)—Anecdotes. 3. Wil-
derness areas—West (U.S.)—Anecdotes. 4. Stegner, Page—
Travel—West (U.S.)—Anecdotes. 5. Rafting (Sports)—West
(U.S.)—Anecdotes. 6. West (U.S.)—Environmental condi-
tions—Anecdotes. 7. Ecology—West (U.S.)—Anecdotes. 8.
Natural history—West (U.S.)—Anecdotes. 9. West (U.S.)—
History—Anecdotes. I. Title.

F595.3.S735 2008
978.—dc22
2007043547

Cover design by Anita van deVen
Interior design by Gopa & Ted2, Inc.
Printed in the United States of America

Counterpoint
2117 Fourth Street
Suite D
Berkeley, CA 94710

www.counterpointpress.com

Distributed by Publishers Group West

10 9 8 7 6 5 4 3 2 1

For Lynn and Allison

TABLE OF CONTENTS

PREFACE

THIS BOOK really begins in July of 1981 when I was invited on an info-trip down the Colorado River with a number of media heavyweights and selected members of Congress. The outing was sponsored by Grand Canyon Dories and the American Wilderness Alliance, and the purpose was to educate lawmakers about the disastrous environmental consequences of a proposal that was then being seriously considered by the Bureau of Reclamation—increasing the power generation capacity of Glen Canyon Dam. A number of big wig conservationists were on board to handle the instructional chores—Roderick Nash, Martin Litton, California's Secretary of Resources, Hughy Johnson. Bill Moyers from PBS was there, as was Robert Jones. The environmental reporter from the *Los Angeles Times*. Tom Brokaw and an NBC film crew joined us by helicopter at Tapites Creek. The assumption was that with all that firepower and assured publicity, the life line of the seventh wonder of the world could surely be saved from the increased destruction planned for it by the feds.

What I was doing in that crowd was unclear. As

a media person, I wasn't even a flyweight contender. As an information resource I didn't know squat about power grids, river flows, riparian destruction, or the reproductive struggles of the humpbacked chub. And I'd never rowed a rapid, or been rowed by anybody else through a rapid, or given a thought to how one went about rowing oneself or somebody elses self through a rapid. But then I stepped out of my car at Lee's Ferry and encountered up close and personal my first *bona fide* river guide, and oh, my Lord, what a Herculean figure that statuary cut, so lean and mean, so monumental, so heroic, six-pack abs and forearms like Westpahlian hams, golden curls atop a soy sauce tan. He strode across the boat ramp in nothing but his wraparound shades, flip-flops and Patagonia Baggers, trailing a faded life jacket behind him with a Gerber River Shorty survival knife affixed to its shoulder pad, pausing now and then to gaze with complete nonchalance out across the roiling waters. That was it. "Stick a fork in me, Mom," I said, "I'm done. Now I know what I want to be when I grow up."

I suffered mightily during that first trip. I hunkered down day after day in the back of a Grand Canyon dory where I hoped nobody could see me sleeping under my Mexican sombrero under a cactus, me, a contemptible tourist, a pathetic, day-tripping, sightseer. I didn't even have a baseball cap. And I resolved I would never, *ever*, be a passive passenger on a river trip again. Two months later I was the proud owner of a 15.5 foot Achilles with 20 inch tubes, Carlisle oars, and a blue Bighorn rowing frame, all purchased from Northwest River Supply in Moscow, Idaho, and shipped to my Santa Cruz door by motor freight. I found a suitably worn life jacket at a rummage sale and strapped a Gerber Shorty to the shoulder pad, and sallied forth to commence my career as an heroic, monumental, Herculean figure (minus the six-pack abs, I regret to say).

The essays that follow chronicle a few of the river adventures, some good and some not so, that I have been lucky to survive over the past 25 years since that initial descent into the madcap unknown. The narrative voice strikes me as a little adolescent at times, a bit cranky and peevish, but for the most part it means well, and I think it may help pass the time while you rest there in your Crazy Creek travel chair under the cottonwoods after a fabulous day on the river, a cold one in hand, waiting for your devilishly handsome river guide with the six-pack abs and soy sauce tan to call you down to the beach for supper.

I should point out to my devoted readership (and you both know who you are) that you will have encountered several of the following narratives in a previous environment. I hope you will not be offended by my indulging in an old authorial tradition, the scheme wherein one first publishes a piece in a magazine, then in an assembly of essays, and finally in one's collected works. It is sometimes disparagingly referred to as "double-dipping" (even triple-dipping), but cash rewards aside, restoration has the irresistible attraction of preserving one's fish wrap for posterity as well as augmenting by a very small number one's imaginary readership. I plead guilty on all counts.

Hence, for those who have an interest in such things, the title essay, "Adios Amigos," first appeared in *Sierra Magazine*, September/October 1995. "Into the Madcap Unknown" is a rewritten composite of three chapters from a book I published with Harper Collins in 1995 entitled *Grand Canyon: The Great Abyss*. "Missouri Breaks" first appeared in the *North Dakota Quarterly* in the fall of 1986 and was reprinted with an additional section on Lewis and Clark in *Sierra Magazine*, June 2000. "Nuestra Señora de los Dolores" was a contribution written in haste for a book I edited for Harcourt Brace in 1996 entitled *Call of the River*. I disliked it very much, and after it was published rewrote it from stem to stern just to satisfy my soul. And finally, "The Bright Edge of the World" is calved from several essays, one called *Red Ledge Province* that appeared in *Sierra Magazine*, March/April 1994, and the other *Canyonlands Expansion* in *New West*, August 1999.

Bon appetit.

ADIOS AMIGOS

THE BUREAU of Land Management (BLM) launch site for the lower section of the Owyhee River is about a quarter mile past Rome on Highway 95 in the extreme southeastern corner of Oregon, easy enough to find if you don't blink going through Rome and if you're not asleep at the wheel. We haven't seen another vehicle since we passed the Fort McDermitt Indian Reservation on the Nevada border, but there is a cute little quarter moon descending on the elongated shadow of Steens Mountain that gives me something to focus on.

I turn on the radio and fiddle with the dial, getting a clear signal from Del Rio, Texas, until we top a low rise and begin a gradual descent toward the river plain. A station from Elko fades in—news and a weather report about a freak storm yesterday that dumped 4 inches of snow in the higher elevations of the Humboldt National Forest. Must be getting time warp here as well. Snow? In May? It was 85 degrees when we stopped in Winnemucca.

Through the rearview mirror I can see into the dark recess of the van where my companions sleep among disarrayed piles of gear. In addition to the personal items strewn all around, we have one 15-foot Achilles raft, one oar frame and three oars, one cooler, two dry boxes of food, a fire-pan, a grate, a 55-millimeter rocket can with toilet seat to serve as a WC, and the collective experience of three prior river trips. All the rowing experience is mine, all of it gained following practiced boatmen through perilous rapids that alone I might have found mystifying. Actually . . . crucifying.

Well, no problem. The Owyhee, according to my research, is a float trip, a mild three- or four-day bump-and-grind down a class II stream more sportingly challenged in a rubber ducky than in a big, heavy-duty, hypalon raft capable of negotiating Grand Canyon–sized white water. According to the BLM river guide, the stretch below Rome we plan to float is "the best choice for family groups and parties with inexperienced members," and the ideal water flow levels are between 1,000 and 4,000 cubic feet per second (cfs). The ranger I spoke with at the river forecast center a few days before we left California indicated a week's average somewhere in the middle of that spectrum. "Things could change," he said. "They say we might even get some snow. Probably not, but I'd have some warm clothes just in case."

Snow? Why are all these idiots talking about snow? The launch site east of the Rome bridge shows no sign that precipitation, frozen or otherwise, has ever fallen on this high intermountain desert, this volcanic wilderness the BLM classifies as a rhyolite canyonlands/

sagebrush-bunchgrass ecosystem where the prong-horn and mule deer play.

In the headlights of the van, the launch site camping facilities advertised in the guide look like a pretty hardpan pallet upon which to lay one's bones, but it's past midnight and I'm so tired I don't care where I sleep. Neither, obviously, do my compadres, since none of them stirs when I cut the engine and go around to the back to dig out my sleeping bag. I lay it out beneath a kindly juniper and stretch out, lulled by the sound of the nearby river rolling in its banks. I do not get in the bag. My last conscious thought is that the air temperature must still be in the 80s. "Snow," I think. "What a joke."

Bud has made a pot of coffee, hauled the deflated raft out of the van, and is blowing it up with the Carlson pump when I wake and crawl out from under my tree. He greets me with a gesture—a backward thumb toward the river I could only hear last night, not see—and offers the opinion that it seems a little higher and faster than he'd imagined a bump-and-grind class II stream would be. "Maybe it's real shallow through here," I say. "There's supposed to be a gauge at the bridge."

"I checked it."

"And?"

"And I didn't believe it so I went to the ranger's trailer over there and talked to him. He said it was 8,000 cfs last night; it's 10,000 this morning."

I consider this news in light of my own expectations and find no place to accommodate it. River flow is relative. Anyway, Bud has always been a bit of an alarmist when confronted by the unforeseen. "Well," I say, "the Mississippi is 650,000, and it didn't worry Huck Finn."

"I don't believe there's any rapids in the Mississippi," Bud says.

"Well, there won't be any rapids in the Owyhee either," I argue. "They'll all be underwater."

"We'll see," Bud says.

My wife, Lynn, loads food into the dry boxes and the cooler while I finish inflating the Achilles and securing the rowing frame. We pack our duffel in waterproof Bill's bags and lash them in the rear compartment of the boat on a cargo sling suspended above the floor, and then Bud takes the van across the bridge to the gas station in Rome and leaves it with the woman who will shuttle it to our take-out point at Leslie Gulch. When he returns he volunteers the information that the gauge by the bridge now reads 11,000 cfs—a 1,000-cubic-feet-per-second increase in less than two hours. "I been looking at that BLM pamphlet," he says, producing the very document all rolled up from his back pocket. "It says here, 'Although the lower section has been safely boated well above 12,000 cfs, extreme caution is advised. Huge holes, standing waves, and fierce hydraulics begin to develop at levels above 7,000 to 8,000 cfs.' What do you think?"

Onward. Charge. Ready, fire, aim. Never apologize; never explain. That's what I think.

Both the main stem and the south fork of the Owyhee begin in a mountain range that lies 100 miles east of here along the Idaho-Nevada border. Tributary streams from the western slope of the Owyhee Mountains contribute to the general flow as it carves its way through 200 miles of switchback canyons to its ultimate demise in the silt pond called Lake Owyhee, but most of its water comes from the Humboldt National Forest drainage north of Elko—Elko, origin of the weather report that ate Del Rio. Origin of those rumors about snow. The same snow that seems to have actually fallen and has now been subjected to 80-degree temperatures for two full days, that is now coursing down every crack and crevice in those headwater mountains, is now trickling, running, gushing, pouring, surging, flooding down every fissure, furrow, ditch, gully, gulch, ravine, and canyon, is overflowing the Wild Horse Dam above Mountain City, is ripping down banks, tearing out trees, washing away livestock, cow and cowgit alike, women and children, little does and lambsy-divey. Well . . . perhaps I'm losing it. Perhaps it's enough to say that an unusually late winter storm dumped a lot of snow that unusually early summer temperatures have been melting. Naturally, it takes a while for the runoff to arrive 150 miles downstream. But the highest flow ever recorded in May— twenty-five years ago, on the first of the month—was

14,000 cfs. And was safely boated too, I feel sure. Of course, at this point we don't know any of the above. But we will, as advised, exercise caution.

The three of us are horsing the heavily loaded raft down to the water when a BLM ranger saunters down to see us off. "Name's Jim," he says, shaking hands all around. "Welcome to the Owyhee." He wears only a pair of canvas shorts, flip-flops, and a green baseball cap sporting the BLM logo on its forepeak. We pump him for information about rapids, and he tells us that except for "Septic Tank" and "Adios Amigos" (names changed to protect the innocent), we're not likely to run into anything too serious. "Septic Tank is a no-brainer," he says, "but Adios Amigos can be a little dicey because it's around a sharp bend in the canyon and you can't see or hear it coming. But hey, this is just a little bump-and-grind river. No problems."

"It says here," Bud observes, holding the spindled guide in one hand and a copy of William McGinnis's book on white-water rafting in the other, "that Adios Amigos is a class III rapid. Class III means 'high, irregular waves, obvious holes and rocks but some-what difficult maneuvering required, advance scouting highly recommended.'"

"Well," Jim says, squirting a stream of Skoal between his flip-flops, "might be a IV in high water."

Bud consults McGinnis. "Class IV: Very diffi-cult. Long rapids; waves powerful, irregular; danger-ous rocks; boiling eddies; passages difficult to scout;

scouting mandatory first time; powerful and precise maneuvering required; demands expert boatman and excellent boat and outfit."

"Well, we've got the excellent boat and outfit," I say, dragging the raft by the flip lines out into the water.

As we pass underneath the bridge I find that, without moving in my seat on the rowing frame, I can reach up and touch the steel girders supporting the road-bed. The watermark on the gauge now reads 12,000 cfs, a fact Bud notes as he mournfully regards the receding highway. Lynn, in contrast, is focused on the half-submerged, totally dead steer that has snuck up alongside the left-front tube of the raft, bloated, stinking, its hide showing like cold grease through the sparse red hairs on its belly. "Oh, puke," she wails, clapping her hand to her mouth.

The main channel is hard to find. The bottomland for a mile below the put-in is so flooded that the rushes and sedges lining the established banks have disappeared beneath the silt-laden water, and when the current splits around a bar or a logjam we have to guess which way lies a grounding, which way a break-through. Eventually the canyon walls ascend and the river is confined to the bed it has been carving for millions of years through the Owyhee Uplands, through the rimrock cap of basalt and interbedded sedimentary units, and down through 1,000 feet of rhyolite to its present depth.

The walls go higher. Where we stop for lunch on

a narrow beach they drop almost vertically, some 300 feet, and there is no escape now but down the river . . . which is continuing to rise. I poke a stick in the sand at the shoreline and retreat to the sandwiches and cold drinks being served on the terrace, underneath a hackberry tree. The rapid flight of a kestrel draws my eye to the canyon rim and then above to the soaring testimony of a red-tailed hawk. When I eventually return my attention to earth and the river, my little stick has been swept away.

Flood-speed carries us downstream at twice the 3- to 4-mile-an-hour pace we'd anticipated, and none of the class II rapids indicated in the guide are extant at this water level. Were it not for an occasional landmark along the bank, we would have no idea of our location, no notion of our progress, no way to guess our position in relation to the upcoming "no-brainer" at Septic Tank. As it turns out, it doesn't matter. We can hear the rapid for almost a mile before we see a nasty incandescence of exploding water where the river slams through a boulder field and churns down through constricting canyon walls, rebounding off one side and then the other, smashing back into itself in a pileup of deranged fury, rushing on through the boiling foam of a gargantuan reversal until it dissipates, finally, in a series of steep-sided standing waves.

We pull off to scout. There are three entrances to the rapid, three "tongues" of slick water leading into the tumult, but it does not take a William McGinnis to figure it out. The first channel leads into an eddy of dead cows; the second into an uprooted cottonwood tree jammed between two rocks. From where we have pulled up to the bank there's no certainty that I can row 2,000 pounds of boat, gear, and frail male and female pulchritude across the river in time to hit the third option, but since failure means wrapping the raft around a colossal fang of rhyolite and dumping the whole shiteree into a raging cauldron, I intend to give it my best shot.

Bud pushes us off and springs for the bow tube as I lay into the oars with all my strength, providing a sudden burst of momentum that causes him to miscalculate his leap and wind up 6 inches shy of his mark—causes him, in short, to fall in the river. He maintains a death grip on the bowline, however, and while I flatten my ferry angle in what proves to be a useless, too-little-too-late attempt to slow our descent, Lynn pulls frantically on his life jacket to try to retrieve him. Panic starts to set in. We're being swept sideways without forward progress. Pull, I tell myself, pull, pull, pull, it's that last stroke between making it and wiping out, got to keep a downstream angle or we are not going to . . . "I got his hat," Lynn yells over the roar of the approaching maelstrom. Cool, something to remember him by. In the corner of my eye I see the boulder field racing toward us and I haul at the sticks in such a frenzy I nearly topple myself out of the boat when the turbulence tilts us 45 degrees and I grab air with my downstream oar.

All of a sudden we are into the smooth water of the

tongue, we're lined up where we want to be ... but Bud is still hanging off the bow line. "Let GO," I bellow at him. "SWIM, damn it." But he only makes another frantic grab for the tube and I'm forced into a desperate, last-second, double-oar pivot. We slide down the V-slick backward to keep from running him over in the mouth of the rapid. Not good, this. Can't see where I'm going. We slam into a curtain of foam and spray thrown up by a submerged rock patch, spin crazily into a lateral wave that half fills the boat, and carom off a pillow on the upstream side of a boulder. The oars are torn out of my hands by powerful hydraulics as we flump into a hole on the downstream side, and before I know it I am slapped off my seat and floundering in a 300-gallon tsunami of dirty river water rushing over the floor of the boat, along with Bud's hat and an empty tuna-fish can. Fifty gallons go into my gaping mouth, and my head smacks the aluminum pipe of the rowing frame when I try to regain my seat. Back on the floor I lose all sense of direction, along with both kneecaps as we are hammered by a submerged rock.

And then things grow strangely calm. I hear cheering topside. "Yeaaa. We made it. Alright." Cautiously I poke my head up and peer over the thwart tube at my jubilant crew, Bud included. "How'd you get back?" I say.

"Don't know," he says. "What goes around, comes around. One minute I'm out, next minute I'm in. No thanks to you, by the way."

"Listen, man, about my yelling at you like that,

most inconsiderate, please accept my . . ." But before the apology leaves my mouth the front of the raft abruptly departs from view, there is a violent shudder from beneath, as if Moby Dick has just rammed the *Pequod*, and then the bow reappears, high overhead, with Bud and Lynn pasted desperately to tubes that have suddenly been folded in half by the force of a back-curling, standing wave. For a moment they stick there like gnats on flypaper; then the old tub miraculously punches through the keeper, the tubes catapult back, and they are popcorned into the river. Highly comedic, this, until the stern whipsnaps over the wave and I am similarly ejected. We all take a short, post-prandial dip, washing out in an eddy 100 yards downstream, along with the boat and a decomposed mule deer. "Nice run," Bud wheezes. "Smooth."

At a part of the canyon where the walls pull back and begin to contour themselves gradually to the plateau above, we stop and make an early camp. Widely scattered juniper dot the slope above the banks, and Wyoming big sage is thick all the way up to the short vertical pitches below the rim. Except for this sporadic vegetation, the landscape looks as if it had been hammered into sulfuric fragments, though where we unload the boat there are some moderately sized cottonwoods providing protection from the late afternoon sun, and the ground beneath them is soft, loamy sand—perfect for reclining against a Bill's bag with a large whiskey and contemplating the events of the day.

Although these canyons are home to bighorn sheep, bobcat, mountain lion, beaver, otter, mule deer, and a host of smaller critters, we have seen little in the wildlife department thus far except *bovinus rigor mortis*, and, of course, birds—kestrels and hawks, ducks, chukar, and Canada geese with their tiny broods of lime-green chicks. It's probably too late in the spring to see bald eagles.

A yell from upstream distracts my attempts to stone the whiptail that has just undertaken a possibly fatal journey up the barkless trunk of a fallen cottonwood, and I look up to see a yellow Maravia raft coasting along the bank. The river ranger, Jim. Just in time for supper. Damn. There goes the extra steak. I take his bowline and tie it off, and he accepts the invitation to a cold can of Old Milwaukee (never waste good liquor on a government official). He has been collecting brand tags off the dead cows floating in the river ("Dumb bastards get trapped against the canyon walls in high water, and just stand there till they drown"), and we engage in a two-beer speculation on just exactly how high the water has actually risen. Clearly, it's still coming up.

"It was 15,000 when I left Rome," the ranger says. "Never seen it like this."

Which inspires Jim to decline our invitation to supper. "I better take advantage of the remaining light," he says. "See if I can get down close to Adios Amigos before things get too much worse."

* * * * *

Which is perhaps why my companions are somewhat subdued as we shove off on the morning of our second day. Neither has much to say. The pair of golden eagles we see shortly after we put in elicit a few moments of communal excitement, but everybody soon submerges into the murk of private thought, coming up only occasionally to check the river guide in a useless attempt to pinpoint our location.

Tiny mosquito-colored clouds begin to appear high above the plateaus; by noon the sky has turned gray and an upstream wind starts to blow. We pull on sweaters. Bud climbs over me and sits in the stern of the raft. Lynn remains in the bow where lateral waves, spanking the front tubes, send an occasional fan of water into her face.

She makes an unladylike comment and suggests that I pay more attention to my rowing. The hag. And what's Bud doing hunkering down back there out of the weather? I don't hear anybody offering to spell me at the oars.

The canyon walls have closed in again and now rise almost vertically from rocky banks against which the detritus-laden current swirls and eddies like batter in a mixing bowl. All bends appear to sweep to the right, all straightaways to end in promontories that create the illusion of a disappearing channel leading into . . . Adios Amigos. We stop and scout everything, climbing phantasmagoric paths to summits that peer down on just another stretch of unmarred river, on

dung-colored water flowing unfettered between deso-
late shores, on the bobbing blemish of a decomposed
steer. No rapid. By the fifth false ascent, I realize I am
growing tired.

It is immediately apparent, when at last we round
the proper corner, why rafting the Owyhee during a
flood is not recommended. Adios Amigos, described
as a mild class III drop during normal levels of 1,000
to 4,000 cfs, boggles the mind above 15,000—staggers
the imagination, defies reason, surpasseth understanding. We do not know its present volume, as we circle
in the eddy, staring at great flumes of water thrown up
by massive boulders that choke the river corridor from
one vertical wall to the other, but we figure it must
now be close to 20,000.

It is a conjecture we will come to remember as a
bit imprecise. We will learn, in time, that today the
Owyhee is cresting about 24,000 cubic feet per second, higher than the average daily flow of the Colorado through the Grand Canyon. We will learn that
we are witnessing the highest runoff ever recorded on
this river. We will learn other things as well, but at the
moment we merely gape and pull for safety.

A thick mat of debris in the eddy makes it difficult
to maneuver the boat, but the rotational current carries
us back upstream along the bank until we reach a place
where we are able to tie off, scramble onto the rocks,
and climb the steep talus to a vantage point above the
rapid. The news is not much improved by the view, but
there is, at least, evidence of a small passage through all

this violence—a passage, unfortunately, that lies clear on the other side of the river. If the fast current (or the condition of the boatman) causes an early entry into the canyon's constricted throat, there is absolutely no question about the outcome. Adios Amigos could flip a Mississippi River barge.

And if one were somehow to survive the initial frenzy, were not dismembered in the ensuing shredder of broken rock, one would find little cause for celebration—a short 100-yard swim in wild, boiling, unmanageable current before being lofted into a 20-foot wave. The wave breaks backward off an undercut wall where the canyon suddenly doglegs to the left; the hapless boater caught in it is guaranteed a pounding ride to the bottom of the river, there to be plastered against the cliff until the reemergence of a quieter, gentler Owyhee in the lower flows of early summer. There is no need for us to hold a conference. One long stare is worth a thousand words.

The look on my passengers' faces suggests that they have little enthusiasm for what we are about to attempt, but they take their battle stations without giving expression to their apprehension. Lynn coils the bowline, tosses it to Bud, and pushes us away from the bank. Assuming things go as planned, the upstream current will carry us to the top of this huge revolving pond where, with a few hard pulls on the oars, we will bust through the eddy fence and execute a mad, downstream ferry across the river.

Things do not go as planned. I find that the mat

of twigs, straw, branches, reeds, and brush trapped in the eddy is at least a foot thick; it is next to impossible to get the oars in or out of the water. And when we do reach the "fence," that crazy line where eddy current and river current collide, it turns out to be ten times more powerful than I anticipated, deflecting us back into the whirly-go-round so abruptly that for a moment I think we've hit a wall. This is disheartening. From some cave in my brain I can hear someone yelling "pull, pull, pull," but we are drifting quickly downstream now, and if I cannot break out at the top of the eddy I will not have time to ferry this fat, overloaded, plastic toad to the other side. And if I am not on the other side and lined up precisely with those foaming jaws through which a passage lies, we are going to become fatality statistics in the annual boating report.

I tug us inshore through the muck so that the current can carry us upstream for a second try, frustration and anger mixing with uncertainty, the muscles in my arms starting to spasm. I wipe at the sweat stinging my eyes and see that I have torn the calluses off both hands. The oar handles are bloody.

When we reach the upper point of the eddy again, I brace my feet against the frame, force the blades through the detritus, and pull with everything I've got. The shafts bend, we lurch into the narrow band of clear water along the eddy fence . . . and are bounced away like a billiard ball off a cushion. "Pull," somebody screams, "pull."

I try once more, but it's no good. There's no strength left in my arms, and my hands feel as if they have been skinned, my fingers filleted. I tell Bud to take the bowline and bail out when we go around again, and we abandon ship with a mixture of relief and regret, collapsing on the rocks to contemplate the cost–benefit ratio of our setback. We haven't died—yet—but we can't stay on this rock pile either. And there's no way to climb out; the walls are 500 feet high.

There is still plenty of light, but the sun has long dropped beyond the western rim of the canyon, leaving us in melancholy shade. We know we can't just sit and ponder, yet both Bud and Lynn seem willing to squat on the boulders and stare at the river. My failure to break through the eddy fence has seriously compromised their expectations of survival. "This is totally the shits," Lynn says, and then fetches some oranges from the dry box and passes them around. "What do you think would happen if we were to tow the boat up above the eddy?" she says. "Start across before things get weird."

The only objection to this idea I can think of is that it might work. It will get us right out there where I can screw up and drown three people.

It takes us an hour, but with one of us on the bowline, one on a stern line, and one in the raft to fend off when we're grounded or hung up in the brush, we get the boat above the sucking grasp of the eddy. It is exhausting, shin-bloodying work. Swarms of mosquitoes feed on us in a euphoric frenzy, we are flayed by thorns and tortured by black gnats, and the light is fading. Time is running short. "We better go for it," Bud says, when we finally stand and catch our breath.

My mouth is an ash pit and I can't reply. I nod dumbly and follow the others down through the rocks to the boat, feeling like the condemned man on the last cold dawn, listening to the footsteps of the warden coming down the hall. I put on my life jacket, untie the bowline and coil it, turn back up the slope to give the world a half-hearted wave, and find myself looking straight into the ice-blue eyes of ranger Jim. "What the hell happened?" he says.

I gawk at him, uncomprehending. "Where did you come from?"

"I got down here last night just before dark, took one look at that eddy, and decided to take a chance; stayed on the left bank and ran it blind."

"Yes," I mutter, "the eddy . . . forever eddy . . ."

"I hung out in camp this morning. Got worried after a while when you never showed up, so I worked my way back along the cliffs. Worse than the damn rapid."

"Rapid. Is there a rapid?"

"If it's okay I'll just snag a ride down with you guys. I'm about half fried."

"Sure," I say, "only thing is . . ." There is a strange lag between sound and cognition, as if speech precedes thought. I can almost see my words float out of my mouth and hang in the air. And I sense a tiny subplot beginning to form in the recesses of my mind

. . . young Atlas, here, deltoids like footballs . . . ah, how fine to be twenty-five and in the peak of health.

What is it I want to impart? There is a precarious sense of pride, dignity, and respect on the line here. Or is it only ego, machismo, and testosterone—supercilious measures of courage and manhood? Am I going to fall victim to the yahoo creed of the white-water river runner? But hey, pride and chemistry goeth before a fall, you know, so let's think this thing out. There are lives besides my own at stake here. I'm in no condition to be doing this. Nevertheless, I seem about to. Or maybe not. Unclear whether I define an expression of responsibility in action or just hormonal impairment. Let us consider the hours put in at the oars, the number of scouting hikes undertaken, the muscle spasms already occurring in the lower back. And there are these raw, bleeding hands . . .

"Jimbo," I say finally, "old buddy. I think you better stick a fork in me because I'm done. You the man."

Jim looks confused. "What?"

"I say, you're the man. To drive the boat."

"The boat?"

"Down the rapido."

"Aw . . . well, the BLM won't let me do that in a non-agency craft. I mean, it's a liability thing. I could lose my job."

"Better your job than your life, right?"

"Well, yeah, but . . ."

"I feel sure the agency will understand."

He gives me a long, unhappy look of . . . who knows

what, compassion? Contempt? I don't care which. "Well, okay," he says, perceiving he has no choice, "okay."

The rest occurs as if in a dream. As we slip through the rapid's maw I imagine an inconceivable wall of water towering above our tiny rag of rubber and glue, a fantasy that hangs over us momentarily like a cartoon disaster, then falls with an implosive force that I fancy must surely rip us in half. The flip lines stretch and cut into my hands as I am pounded to the floor and dragged backward under the thwart tube. The raft is swamped. I fight to free myself, to pull myself up and grasp a lungful of air, and as I do I see a representation of Jim, still upright, still with his feet jammed for support under the tubing of the frame, the tendons in his neck standing out like guy wires to hold his head on as he strains against an inestimable force. Then I am turned sideways and spun around. I see a gray wall of rock rushing toward us with a great, chocolate wave looping backward. Déjà vu, I think. A vision of carnage and death. Adios Amigos, it's been a great trip, a bit unusual, but the Côtes du Rhône was superb with the fillet and fried potatoes, sorry we didn't see a bald eagle . . .

Fortunately, nothing is forever, even the interminable wait for the final curtain. But suddenly the pounding stops, and it feels so good it takes a while to realize that I am in a real raft full of real water, upright, not in the river where I ought to be, plastered against an undercut wall for the suckers and catfish to pick at until spring subsidence; I am on all fours on the flexible hypalon floor, bobbing up and down and looking over the front tubes at the beach where we are about to land. Then I feel the sand grind under me and my knees on terra firma, and watch stupidly as Jim leaps out, takes three steps and falls flat on his back, legs rigid with muscle spasms, heels beating a tattoo against the sand.

Then I flounder out of the raft too, and Bud and I hold his feet down while Lynn pounds on his calves. Gradually the cramping stops, the quivering subsides, and we all sit silently, staring at the river. After a while Bud commences pouring the water out of an ammo can that contains his sunglasses, ChapStick, chewing gum, canister of whoopee weed, and a short descriptive pamphlet on floating the Owyhee—which he picks up and, after a moment's perusal, reads to us aloud: "A mild class II excursion from the Rome bridge to Lake Owyhee Reservoir; suitable for small rafts, sportyaks, and rubber duckies; excellent stream for beginning boaters." He wads the pamphlet into a ball and throws it out into the current. "I think next time we should run a river with some starch to it," he says, gazing back toward the bend in the canyon from which we have been ejected. "You know, something more dramatic than these wimpy little bump-and-grind rivulets, like, say, Niagara Falls . . . in a canoe."

We take Jim to his raft a half mile farther down the river and make camp for the night. In the morning we are up at first light, so eager to get done with this river that we hardly bother with breakfast. We wish Jim luck collecting his brand tags and shove off in a gray dawn that never seems to break, thankful for the fact that there is nothing between us and Lake Owyhee now but high, flat water. Even the weir Jim warns us about is washed out.

By ten o'clock we begin to lose the current, by eleven we are rowing through the marshlands at the upper end of the lake, and shortly thereafter we emerge onto the flat, leaden sheet of the reservoir, a colorless expanse only distinguishable from the countryside beyond by a thin bank of mud along the shore. Pallid sky merges with bleached grasslands in a seamless union, and here and there, dotting the lake like little dumplings in a kettle of sediment stew, are the swollen carcasses of scores of dead cows, flushed into the settling pond by the bombastic cataract above.

Ten miles to cross this sullen sea. We make slow progress. After a while time loses its reality, and, for that matter, so do we. We are like rub-a-dub-dub, three characters in a tub. We are just rowers in a monochromatic sequence of absurd conjunctions. Clearly this whole episode has been excerpted from some crazy Italian-Swedish coproduction by Federico Fellini and Ingmar Bergman. It certainly has no place in this panegyric to the joys of whitewater rafting. And I wouldn't put it in any promo guide to the lower Owyhee either.

INTO THE
MADCAP UNKNOWN

September 18

IT IS EARLY FALL, and high up on the Kaibab Plateau a few stands of aspen have jumped the gun and are starting to show tints of yellow. An autumn breeze blows through the spruce and fir, and the red squirrels are going frantically about their last-minute winter storage. But down here on the Marble Platform, the air temperature is in the upper 80s, the rocks along the Colorado River bank are hot enough to fry a rasher of bacon, and the water is 48 degrees. Standing in it for more than twenty seconds makes the legs burn with the cold.

There are fourteen of us at Lee's Ferry, blowing up the boats, tying down raft frames and oars, loading the waterproof duffel bags, rocket cans, ammo cans, dry boxes, water containers, fire pans, medical kits, repair kits, pumps, spare life jackets, and the twenty-four cases of beer needed for barter and trade. Commercial operators hog the ramp with their 30-foot baloney boats and their incredible mountains of gear, but we will endure their snarling outboard engines for only a few days as they pass us going down the river. Then their season is over. Only oar-powered rafts are permitted after September 24.

It is nearly dark by the time we are rigged, and we head for the trucks, drive up the narrow access road that leads down from Highway 89 to the confluence of the Colorado and Paria River (or the Pah Rhear River, as George Bradley, one of the boatman with the 1869 Powell Expedition, called it). We take our places sedately in the dining room of the Marble Canyon Lodge at the foot of the Vermilion Cliffs. Order up the Last Supper. There is great merriment among my compadres, and a plethora of bad boatmen jokes, but I feel like Major John Wesley Powell when he wrote, "We are now ready to start on our way down the Great Unknown." His companions remained cheerful, he said: "Jests are bandied about freely this morning; but to me the cheer is somber and the jests are ghastly."

It has taken my wife, Lynn, nine years of annually renewing herself on a National Park Service waiting list to secure a private permit to run the Colorado through the Grand Canyon, and in that time I have gone from a lean, mean, forty-seven-year-old light-heavyweight to an oleaginous, fifty-six-year-old cruiser-weight with bad nerves. Although I've only had one serious wreck in fifteen years of river running (flipping on the Dolores River in the particularly poisonous and mean-spirited rapid called Snaggletooth), the Colorado has many Snaggleteeth on its menu, plus a few dozen rotten molars, and my bad dreams on the tarmac at Lee's Ferry are of maelstroms like Unkar, Hance, Sockdolager, Grapevine, Horn, Hermit, Granite, Crystal, Deubendorff, and Lava. Martin Litton, the founder and former owner of Grand Canyon Dories, used to comfort anxious passengers who asked about the severity of any given rapid by rolling his eyes and intoning, "Horrible. Just terrifying."

Yes, well, Martin has been down this river so many times he can afford to be whimsical. For the rest of us, the Colorado's rapids are ranked on a scale of difficulty from class I to class X, and there are, by my count (and depending on the water level), about twenty-five that are ranked VI or above, fourteen rated VII or above, nine judged to be VIII or above, and then there are Crystal and Lava Falls. For some reason Crystal is the one that has truly captured my imagination, no doubt because after its rearrangement by the 1983 flood it has become the most dangerous in which to make an error. On the other hand, any rapid is dangerous when you make the right error. Horrible. Terrifying.

The National Park Service assigns about 90 percent of the user days on the river to commercial operators, a policy that may have seemed reasonable in the 1980s when few people possessed the skill or equipment to run the canyon, but today it can only be regarded as a serious overallocation of a public resource to private enterprise. So I hold the river ranger's office personally responsible for the recent decline in my eagerness to row class X rapids. I just got too old and fat while

I waited. And from the look of things going into the huge Gott coolers we carry with us, I'm not going to get any skinnier on this trip either. The problem is how to get it all stuffed onto four 18-foot rafts and still have room for passengers.

Which happens to present us with our first interpersonal friction. We have among us a young boatman whose culinary persuasion leans toward seeds and nuts and whose muttered opinion of our floating restaurant and bar is perhaps less than generous. He isn't happy with the boat he's stuck with either, an old Havasu with warped oars, but he lacks seniority and is pretty much ignored. Other differences will arise as we descend the "Great Unknown," but for the moment we are still one big blissful family.

September 20

The Colorado River drops 10,000 feet from its headwaters in the Wind River Range of Wyoming to its outlet in the Gulf of California (actually, it never quite makes it all the way, dying in the sands of Laguna Salada a few miles short of the gulf). In total, it is about 1,700 miles long and drains nearly 250,000 square miles—an area encompassing a significant portion of seven western states. Its Grand Canyon section, between Lee's Ferry and the Grand Wash Cliffs, is 277 miles long—though the last 40 miles now lie beneath the waters of Lake Mead. The canyon area contained within the national park itself encompasses about 1,892 square miles, changing in depth from less than 1,000 feet in the upper sections of Marble Canyon to 6,000 at its deepest point in Granite Gorge. It varies in width from less than a mile to 17.5 miles from rim to rim.

It has been praised as the most "sublime spectacle" in the world and condemned as the most "profitless locality" on earth. Some have peered into it with a sense of oppression and horror; others have been moved to great acts of creative expression and a sense of profound spiritual identification. It has inspired agony and ecstasy in thousands of photographers trying to deal with its changing light; it has moved poets and writers to absolute excesses of purple (and sometimes indecipherable) prose. There is, after all, nothing quite like this canyon anywhere else on the face of the earth.

We can only speculate what the first white men to encounter the Grand Canyon thought about their discovery. None of them got exactly lyrical. A group of Coronado's men under Don García López de Cárdenas reached the South Rim in 1540, where they spent three days looking for a way to descend to the river. Unable to get more than a third of the way down and low on water, they retraced their steps south to the Hopi villages whence they came. Having nothing remotely analogous in their experience against which to make comparisons, they reported that some of the rocks in the canyon were "bigger than the great tower of Seville."

Over two hundred years later, in 1776, Fray Francisco Garces, a Franciscan missionary and colleague of Juan Bautista de Anza, strayed into the region of the Aubrey Cliffs, where he met a band of Havasupai and spent an unsuccessful week trying to convert them at their village in Cataract Creek. Garces stolidly described the canyon as "profound" and remarked in his diary that he was "astonished by the roughness of this country, and at the barrier which nature has fixed therein." Beyond that he seemed to have little to say. That same year the famous Dominguez-Escalante party, trying to return to Santa Fe from the Great Basin, wandered around lost for several weeks in a maze of side canyons before eventually discovering a way to cross the Colorado, near the present site of Lee's Ferry. Escalante's account of this ordeal, like those of his predecessors, wastes little time in florid admiration of the scenery.

During the first half of the nineteenth century, a small number of Americans penetrated the canyon region of southern Utah and northern Arizona—fur trappers like William Ashley and James Ohio Pattie, and the Mormon colonizer Jacob Hamblin, who was sent out during the 1850s by Brigham Young to establish settlements at Moab, Lee's Ferry, and St. George. But no systematic exploration of the canyon of the Colorado occurred until 1857, when the U.S. War Department ordered Lieutenant Joseph Christmas Ives to attempt to navigate the river from its mouth near Fort Yuma to the Mormon settlements in Utah. A 58-foot steel-

hulled steamboat was built in Philadelphia, disman-
tled and shipped around the horn to San Francisco,
carried overland in wagons to the Gulf of California,
and reassembled. On the eleventh of January, Ives and
a company of twenty-four men departed Fort Yuma
and steamed north in the newly christened *Explorer*.
Almost immediately they ran aground.

It was a bad omen. For the next 150 miles, the steamboat repeatedly encountered shoal waters, sunken
rocks, and rapids. It ran aground a half dozen times,
and then finally suffered a horrendous wreck that cat-
apulted everybody near the bow into the drink and
nearly tore the stem out of the boat. Ives realized that
he had gone as far by water as he was going to go, and
if the mission was to succeed, it would have to continue on foot.

The company marched north and then east as far
as Cataract Canyon (now called Havasu) and the village of the Havasupai Indians—a point that on today's
maps would lie roughly midway between Lake Mead
and Lake Powell and nearly opposite the great canyon of Kanab Creek. Ives mistakenly identified Kanab
Canyon for the main branch of the Colorado and,
being unable to cross the river and venture into it,
decided he had now gone as far as he was going to go,
period, and turned back.

Not that it seemed to make him all that unhappy.
From the tone of voice evident in his 1861 *Report upon
the Colorado River of the West*, Lieutenant Ives appears
to have seen enough. "The region last explored is, of

course, altogether valueless. It can be approached only from the south, and after entering it there is nothing to do but to leave. Ours has been the first, and will doubtless be the last, party of whites to visit this profitless locality. It seems intended by nature that the Colorado River, along the greater portion of its lonely and majestic way, shall be forever unvisited and undisturbed."

Our camp tonight is at lonely and majestic mile 33 1/2, just below Red Wall Cavern. We have thus far survived a Native American uprising at Navajo Bridge; numerous class V river uprisings at Badger Creek, Soap Creek, House Rock, North Canyon, 24 1/2 mile, and 25 mile rapids; and a Weight Watchers dinner on our first night featuring turkey with stuffing, mashed potatoes, cranberry sauce, broccoli, tossed salad, and chocolate pie.

The bridge incident (4 miles below Lee's Ferry a couple of Indian kids depth-bombed us with small chunks of sandstone as we floated 470 feet underneath them) reminded me of standing on that same span nearly fifty years ago with my parents and a bunch of people gathered at Marble Canyon for a Navajo rodeo. It was after dark, and my father (among others) was entertaining himself by dumping boxes of Diamond matches over the edge, while Art Greene, owner of the Marble Canyon Lodge, filled a truck tire with gasoline. The matches fell headfirst, of course, and after an interminable wait, ignited in a showery sparkle on the rocks below. The truck tire, when lit and hurled over the railing, descended into the blackened void like a fiery comet in space, landed in the current with a terrific *whomp*, and floated along like Cleopatra's funeral barge until it disappeared around the bend at Six-Mile Wash. Things were better before environmental consciousness spoiled all the fun.

The river through the canyon now descends a total of 1,900 feet, or approximately 7.8 feet per mile, and its flow, controlled by Glen Canyon Dam, currently varies between a high of 20,000 cubic feet per second (cfs) and a low of 5,000 cfs, carrying a canyon-cutting sediment load of about 40,000 tons a day—which, compared to its pre-dam burden of 380,000 to 500,000 tons per day, is like scratching a trench with your fingernail. I have to observe that this discounted trickle has provoked no inordinate challenge thus far, particularly since I have positioned myself about 2 inches off boatman extraordinaire Jim Slade's stern through every rapid we have encountered. Or if Slade's boat isn't available, Ann Cassidy's. Slade and Cassidy have been guides for OARS and Sobek International for years, and collectively they've been down the Grand Canyon about 160 times. I figure if they don't know where they're going, there's certainly no point in my trying to figure it out. In this manner river rafting is greatly simplified.

Below Soap Creek we note the obituary carved in the sandstone for Frank M. Brown, president of the Denver, Colorado Canyon, and Pacific Railroad, who thought he could build a railroad line through the Grand Canyon to haul coal from Colorado to San Diego, and who was of the equally foolish opinion that life jackets were a pointless nuisance on a surveying trip. He was wrong on both counts, and drowned at about mile 12 after he capsized in Soap Creek Rapid.

We doff our hats at mile 24 1/2 rapid to Bert Loper, the grand old man of the Colorado, who drowned after he capsized here in 1949—at the age of almost eighty. And once again at mile 25 rapid to Peter Hansbrough and an unnamed companion who were part of Brown's railroad survey party. Both drowned in 1889, a few days after Hansbrough carved the inscription to his boss below Soap Creek. Horrifying. Terrible.

I cannot say much for my own copycat runs through either of the above riffles, except that I am alive, in camp, and bellied up to the trough when the pork chops, yams, broccoli, lingonberry sauce, coleslaw, and pineapple upside-down cake are being served. Our young boatman has shirked kitchen duty in favor of a photography hike (curious how the light is suitable only when there's work to be done), so we eat his portion and leave him the dishes. First law of the river, "you snooze, you lose," has broad application out here in the wilderness.

September 22

We are now 10 miles below the confluence of the Little Colorado and at the entrance to Unkar Rapid.

The river is making its big westerly bend and is about to forsake Marble Canyon for the constricted walls of Granite Gorge. It will be nearly 100 miles before things begin to open up again, but at least we know what we're in for—more or less. I am reminded once again of Major Powell's over-quoted dirge—recorded, as a matter of fact, at this exact location. "We have an unknown distance yet to run; an unknown river yet to explore. What falls there are, we know not; what rocks beset the channel, we know not; what walls rise over the river, we know not. Ah well! we may conjecture many things."

The modern rafter's problem is that there are not enough things left to conjecture—around every bend lies somebody's favorite side canyon, hike, historical site, archaeological site, geological feature—and what with all the requisite stops at Vesey's Paradise, Redwall Cavern, Buck Farm Canyon, Nankoweap Canyon, and the Little Colorado (or LC, as it is affectionately known), we could well run out of rations before we get to Phantom Ranch. We killed two hours rolling around in the muck at the mouth of the LC, and then the wind came up and started blowing us back up the river. Rowing a fully loaded, 2,000-pound, flat-bottomed, rubber tub against a headwind is a great way to ruin the day.

Neither George Bradley nor Jack Sumner of the Powell expedition thought much of the Little Colorado (formerly the Flax, and before that the Chiquito River)—"a loathsome little stream, so filthy and muddy that it fairly stinks" and "as disgusting a stream as there is on the continent; 3 rods wide and 3 ft. deep, half of its volume and ²/₃ of its weight is mud and silt." But without the Little Colorado, the Colorado itself would need renaming. The sediment load that used to make it run red is now held back by Glen Canyon Dam, and it is only the "loathsome" little stream that reintroduces enough muck to give it back its proper color.

Bradley and Sumner thought the spring bursting out of the Redwall cliff at Vesey's, with its great beard of mosses and ferns and flowering plants, one of the most beautiful sights they had ever seen. (And indeed it is a verdant, brilliant contrast to the monochromatic setting—red rocks/red river—that surrounds it.) Had they understood that it is fed by water sources that seep into the ground a mile up on the Kaibab Plateau, they might also have thought it one of the most amazing.

About the Anasazi ruins high above the river at Nankoweap Canyon the Powell journals say nothing. Probably everybody was too busy with the rapid created by Nankoweap Creek to look up 1,000 feet on the east-facing wall and see the little square doorways of these eight-hundred-year-old granaries. Even though carbon-14 testing of figurines found in a number of archaeological sites indicates that the Grand Canyon was inhabited as far back as four thousand years ago, the Pueblo period of Anasazi culture (of which these granaries are a part) dates from AD 750 to about

AD 1100. By 1150, most likely as the consequence of a prolonged drought, Nankoweap, along with all other sites in the Grand Canyon, had been abandoned. Now it is one of the premier photo-op day hikes for all river trips and always invokes the same question from each sweat-drenched, heart-pounding hiker who has just made the long, steep climb from the river to inspect it. "I wonder why the effing Indians built it so damn far up."

It is possible to get the impression while day-hiking in the canyon that there isn't much out there in the way of animal life—nothing but lizards, ravens, and the abominable *Humanus mucosus*, also known as the great-nosed sightseer. In small part, we think that because few of us are actually in the habit of seeing what we look at; in large part, a significant percentage of the animal life out there is nocturnal. Nothing with an IQ higher than a chuckwalla is going to hang out in the furnace glow of a midsummer afternoon on the South Rim or the Tonto Platform or the river corridor. The maximum mean daily temperature at Phantom Ranch in July is 106 degrees; in the Inner Gorge it can often reach above 115.

The Grand Canyon region actually harbors a huge variety of life—seventy species of mammals, 280 species of birds (of which forty are year-round residents), forty-four species of amphibians and reptiles, and seven species of fish. And, of course, we must not forget everybody's favorite Lower Sonoran creature-feature, the humble scorpion. In the Grand Canyon there are six species of this arachnid, but only the skinny little straw-colored variety known as the "slender scorpion" can seriously ruin one's night. Like most creatures in the canyon, they are nocturnal, and they are prone to scuttle under a tarp, or into a shoe, at the first sign of daylight. Recognition is not always easy, but their consistent yellow color, narrow pincers, and the oblong segments of their very svelte tail (1/8 th inch, maximum) are a help. If, when you find one skulking in your socks, you remain in doubt, you can always poke it with your pinky. And if, after it has stung you, your finger gets numb, and your arm begins to tingle as though you'd slept on it, and you begin to drool, and have trouble talking and breathing, and go into convulsions, and lose your sense of sight—well, congratulations, you've made a positive identification.

A glance at my handy waterproof edition of the Buzz Belknap *Grand Canyon River Guide* (already soaked and its pages stuck together) tells me that tomorrow, in addition to the not-so-gentle swells at Unkar and Nevills Rapids, we have the dubious pleasure of rowing Hance, Sockdolager, Grapevine, Horn, and Granite. We are not talking here about dancing waters and modest sand waves. These are surging monsters with huge back waves, lateral waves, V-waves, tsunami waves, reversals, rocks, jagged neoprene-ripping cliffs, whirlpools, boulder gardens, and boat-eating holes. The wretched hydraulics through these rapids can rip the oar from an oarsman's hands (usually with his arm still attached) as easily as King Kong tears out trees in the forest. These are Godzilla-meets-Bambi rapids. Dreadful. Horrible. Not suitable for senior boatmen.

None of these thoughts are conducive to a good night's sleep, and as I lie in my sleeping bag staring up at what appear to be car lights on the rim near Desert View (the last point before exiting the eastern end of the park), I find the chili relleno casserole, chicken taco, tossed green salad, refritos, and strawberry shortcake sitting a little heavy on the stomach. I think of the Anasazi and wonder if maybe they didn't have the right idea—get out before it's too late. Who would know if I just snuck upstream to Tanner Canyon and high-tailed it up the trail that comes down from Lipan Point? My companions would just think I'd been wandering around in the dark and fell in the river. Swept away. Drowned. Alas, poor Yorick, and all that. Ah well! we may conjecture many things.

SEPTEMBER 24

We have a short river day today, since yesterday was a long pull during which we stopped to scout five rapids and kill a few hours in the delicious, leafy shade of the cottonwoods at Phantom Ranch. Even in late September the temperature on the river gets up into the 90s; anything green provides respite from the heat and glare on the water. The Kaibab and Bright Angel trails converge here, and in the warm afternoon air there is the heady aroma of road apples from the many mule trains that pass through. There are also ruddy-cheeked German girls with walking sticks and short shorts, and French girls, English girls, Dutch girls, Scandinavian girls. And ice cream in the canteen. English is still spoken, some.

I can actually remember the first time I looked upon this bucolic scene. It was nearly fifty years ago, and from the top down. My parents, after a six-year sojourn among the diminutive hills and hummocks of New England, had packed up their worldly possessions; tied them all on top of our Ford station wagon; and were moving lock, stock, and barrel from Massachusetts to California. They took the southern route, old Highway 66, through St. Louis to Oklahoma City, across the Texas panhandle, and into New Mexico and Arizona.

I was six or seven at the time, and half a century has cleared my memory of much of that midsummer odyssey, though a few imagistic scraps and one king-sized revelation still remain. For scraps there is the overloaded, overheating Ford station wagon in which we motored sedately away from the lush greenery of the eastern United States and into the parched, uncharted wastelands beyond the hundredth meridian. A leaking canvas water bag hangs from the hood ornament and a cylindrical "air cooler" attaches to the driver's seat window like a food tray at a drive-in restaurant. It fails to perform any cooling, though it does humidify—a particularly jolly feature through the damp heat of the East and Midwest.

For revelation there is . . . what the British novelist J. B. Priestly called the Grand Canyon—"not a show place, a beauty spot, but a revelation," a place indescribable in "pigments or words." He said he had heard rumors that some were disappointed by the canyon, but he opined that such people would be disappointed by the Day of Judgment. "In fact," he mused, "the Grand Canyon *is* a sort of landscape Day of Judgment" (italics mine).

My revelation was somewhat different from the one experienced by Mr. Priestly, and perhaps it might be worth stepping back in time here for a moment to review it. I should like to film this retrospective from an elevated perspective, however. I no longer wish to be too closely identified with its author.

Let us imagine we are looking down into the rear seat of a Ford station wagon as it travels along Highway 66 east of Tucumcari, New Mexico, circa 1944. A snuffling pre-adolescent is expressing his lack of enthusiasm for cross-country expeditions in a tedious litany of toneless, monosyllabic questions: "Are we there yet? When are we gonna be there? How much farther?"

Trying to cheer the lout, his father begins to tell him fanciful tales about the wild and woolly west, promising encounters with cowboys and Indians, rattlesnakes, scorpions, Gila monsters, and the most horrible creature in the Northern Hemisphere, the "so terrible to look on it gives me the fantods just to think about" saber-toothed jackalope.

No response.

"This thing is all teeth and hair. It can run 80 miles an hour and jump 40 feet in the air, and it devours side-hill cowgits in a single bite. Of course, the cowgit is somewhat at a disadvantage, what with its uphill legs being shorter and all—fine for grazing on steep slopes, but hell on flat ground."

"How far is it?"

"When we get to Arizona."

The desert rolls on, rises slowly to low juniper-piñon woodland, then to pine forest as they climb toward the San Francisco peaks. Dim recollection here of fry bread, or maybe a Navajo taco, then more desert, and father's voice saying, "Well, this is it, bub, jackalope country." Car stopping. Young Fauntleroy, groggy from his postprandial nap in the back seat, told to climb out. Peevishly complies. Is led stumbling across an asphalt parking lot past a sign announcing Yavapai Point. Hands placed on iron pipe railing at the edge of a rocky precipice. Where he stands blinking out across . . . the revelation.

Revelation? Flat rocks, red ledges, a void, an emptiness, nothing, an ensemble of dumb space and fractured horizons, hazy silence, collapsed perspectives. (I'm afraid we may have here one of those people about whom Mr. Priestly heard rumors.) He looks up and down, left and right, scans from rim to layered rim, begins to understand that he's been duped, bamboozled, bilked, gypped, swindled. He's been had. Sputtering disappointment, he turns and wails, "Where's the jackalopes?"

Now it's ruddy-cheeked girls and short shorts. That's bucolic. The not-so-bucolic is that on the way up from the beach to the National Park Service buildings at the outwash of Bright Angel Canyon, there was a notice tacked on a post asking river runners to be on the lookout for the body of a drowned teenager some-where between the Kaibab suspension bridge and Hermit Creek. The young man decided to swim across the river, apparently oblivious to the power of the cur-rent and the rapid loss of motor function in 48-degree water. He was, moreover, of the Frank M. Brown per-suasion when it came to flotation devices, and tried it without his life jacket. A costly error.

The posted notice was not good news—not for me, and not for the patient reader who is wondering why this narrative is fixated (apart from gluttony) on death and dying. In the Grand Canyon the forces of nature are so obviously indifferent to all forms of two-legged life, craven cowards and stud muffins alike, that one's fragility and vulnerability is impossible to ignore. It is made manifest hour by hour, mile by mile, paragraph by paragraph.

Take last night. To the distant roll of thunder we camped somewhere in the vicinity of 94 Mile Creek and ate supper (curried chicken with raisins, sliced apples, Spanish peanuts, shredded coconut, chutney). We did dishes to a symphony of celestial booms and cracks somewhere up on the Kaibab Plateau; we laid out tarps and tents under the flash of strobe lightning and a bombardment of deafening explosions directly over our heads, amplified to ear-splitting volume by the sheer canyon walls.

Lynn and I huddled inside our Moss tent, a delicate little shelter of rip-stop nylon about as airy and insubstantial as Lady Astor's panties. We could hear waterfalls beginning to pour over the ledges a thousand feet above, gigantic cascades of soupy liquid, God's own mud slurry gushing from the Tonto Platform and the North Rim, flowing down over the margins, washing another billion cubic yards of the Colorado Plateau down into Lake Mead. We could hear the roar of multiple falls, we could almost perceive them, in a subliminal, snuff-flick-blip sort of way, but we couldn't tell whether they were actually right here, about to blast us off our tiny little beach into the foaming torrent, or merely distant figments of our terrified imaginations. Eventually the storm passed, and we sank into our sleeping bags, thankful to be dry and alive.

At dawn we awakened to the croak of ravens high on the wall behind our tent, the descending tremolo of canyon wrens, the quiet murmur of the river. We staggered out to discover . . . a brand new gorge cut through the center of our compound. Not exactly another Grand Canyon, but a gaping ravine some 12 feet wide and 10 feet deep that neatly divided our tent site from the kitchen area, and that had funneled away about a third of all the horizontal real estate between the river and the cliff. Had it made its midnight course a few degrees to the right of center some of us would not have been around this morning to enjoy the sausage and eggs, hash browns, cantaloupe, English muffins, blackberry preserves, and steaming pots of cowboy coffee.

SEPTEMBER 25

For thrills and chills today we have Crystal Rapid. Negotiating Crystal is an odious prospect not only because it is dangerous, but because in recent years it has featured so largely in every boatman's favorite pastime—the telling of river horror stories. In 1966 a flash flood in Crystal Creek washed huge boulders down into the Colorado River and created what was, by most standards, a nasty rapid. Then in 1983—the year of the floods, the year Lake Powell lapped at the top of the concrete dam impounding it—the Bureau of Reclamation opened everything it could open at Glen Canyon, and the river ran five times its normal flow, day and night, for weeks. A thousand cubic feet per second. There were serious consequences for the beaches and plant life along the riparian corridor; there were serious consequences at Crystal Rapid. It got rearranged once again. It got worse than nasty.

People who ran Crystal during the flood like to reminisce (perhaps with appropriate embellishment) about the humongous hole created just below Slate Creek, a crater 100 feet wide and 30 feet deep, with a 20-foot standing wall of water at the bottom and a hydraulic force equivalent to tide rips in the Johnston Straits; they like to recall the hapless ones, the massive pontoon rigs that didn't stop to scout the rapid because they didn't know it had been altered, and anyway, with their out-riggers and big motors they could power through any-thing—except that this time they couldn't and went in like breaching whales, spy-hopping in the standing wave and flipping over like plastic toys in a kid's bath-tub. They manufactured some of the most spectacular wrecks ever witnessed, with twenty or thirty people per boat ejected into the icy maelstrom and gear strewn all over the river, food, garbage, port-a-pots, and all.

At least one person drowned, a lot got stranded on rocks and up against sheer cliff walls, and even-tually the Department of the Interior had to bring in Navy Seals to pull people out. The only light moment occurred later during a National Park Service inquiry into the whole catastrophe, when Georgie White, one of the pioneer commercial outfitters on the river, is reported to have observed about her own loss of cli-entele in Crystal, "They just don't make passengers like they used to. I tol''em to hang on."

But that was 1983, the flood year; this is 1993, and Crystal has been somewhat subdued by the regulated flows instituted in 1991. Those flows will continue at least until environmental studies wrap up on the impact of Glen Canyon Dam on the river's resources. From the boulder where I sit staring for about an hour at the Colorado's worst rapid, it still looks like a nightmare, with two bodacious declivities in its carotid artery in which, it seems to me, one could lose a 10-ton truck. I go back up to the boats cotton-mouthed and sniveling and eviscerate myself cinching down my life jacket. There is no way out of this.

Jim Slade leads off; ours is the second boat in line. Like Slade, I go into the top of the rapid along the right bank (but not so far right that I can hit the rocks poking up along the shore and get skewed around), power into the V-wave with a hard downstream ferry, and, in order to stay right—that is, stay away from those raging maws that will certainly swallow me as quickly as a toad swallows a fly—begin straining at the oars so maniacally I think my eyeballs may herniate ... and slide right past both holes. Park in an eddy. Nothing to it. A piece of cake. Never got *damp* even. Don't know why I gave all this a moment's notice. Apologies for my theatrics. I grovel in mortification.

SEPTEMBER 27

This morning Ann Cassidy offered me the casual observation that standing around the kitchen in bare feet was a risky proposition because our camp was heavily infested by harvester ants. Sadly, I ignored her warning. Desert campers and river runners call these cheery little pets fire ants, but not because of their color. A seed-stashing formicidae that lays up great supplies of food in caches 10 feet or more underground—and that has survived the hottest and most arid conditions for millions upon millions of years—it establishes itself in massive colonies generated by a single queen, who mates but once and goes on producing eggs for up to two decades. It is apparently the responsibility of her multitudinous offspring to hang around campsites frequented by crumb-dropping *Homo sapiens*, waiting for an opportunity to crawl between a flip-flop and a big toe and register a presence. In this case, my big toe.

The sting of a harvester ant is an elevating experience, one that is way out of proportion to the size of its perpetrator, one that can leave its recipient massively indifferent to everything but the eradication of ants in general, family, genus, and species—not to mention the single, felonious individual. The bite of a fire ant results in a surfeit of vulgar language, the flinging of the quisling flip-flops into the river, and the reintroduction of shoes and socks. In truth, the only thing positive to be said about these pandemic little buggers is that they are diurnal and go to bed early.

We have passed from Upper Granite Gorge into Middle Granite Gorge and the Granite Narrows. The principal feature here, not surprisingly, is granite, and it can become very oppressive, even when one is cheered by the prospect of another meal. Powell's men didn't like it from the beginning, Jack Sumner reporting just a few miles below Bright Angel Creek that "this part of the canyon is probably the worst hole in America, if not in the world. The gloomy black rocks of the Archean formation drive all the spirit out of a man."

I'm sure it is environmentally incorrect to find fault with the geological splendors of the Grand Canyon, but there are times on this journey when I find the labyrinthine walls as oppressive as did Sumner (who had no idea where they led), and the cramped, confining gorge conveys little more to me than a chilly reminder of my own mortality. The view from the river is not so jolly as the panoramic vistas from the rim, and I often feel as if I'm at the bottom of the grave looking up. Every now and then a cold draft blows across my sunburned shoulders. Chilled, I look around. Out of what dark cleft in these walls of sheer, black granite might that icy finger have come? In the ancient, two-billion-year-old schist I make gloomy comparisons to the nanosecond of my own life.

Two-billion-year-old schist? That is a span of eternity that is completely meaningless, like the information that the earth itself is 4.6 billion years old, or that there are, on average, 100 billion solar masses in a galaxy, or that light traveling in a vacuum for a year covers 9.46 trillion kilometers. Unimaginable. Unthinkable. Preposterous.

Nevertheless, to accommodate those of us of limited intelligence, there have been numerous attempts to conceptualize geological time as it is represented in the Grand Canyon rocks—ancient rocks in what seems to be a very young canyon. One favored gambit is to represent on a twenty-four-hour clock the interval of time between the actual formation of the rock (schist) exposed at the bottom of the Inner Gorge and the beginning of the Colorado River's canyon cutting. In this scheme the world (4.6 billion years old) is formed just after midnight and the schist (1.7 billion years old) in the early afternoon. The river starts carving the canyon (5.5 million years old) just before midnight—at 11:58 PM. Human life begins at 11:59 PM; your personal tenure on earth, dear reader, passes by in less than half a second. None of that is of much comfort as I sit here trying to assimilate what I am looking at, but it may give a clue as to why it makes me feel queasy and insignificant.

If you don't know where you are, Wendell Berry has reminded us, it's difficult to know who you are. To which I might add: in the canyonlands of southern Utah and northern Arizona it's hard to know where you are unless you know where you would have been if you'd been around two billion years ago. It's very weird out here. It requires explanation. And explanations require not only information but leaps of faith and lots of imagination.

It is easier to try to comprehend the Grand Canyon from almost any elevated point along the 10,700 square miles encompassing it than from down here at the bottom of the well. The first step is to ignore the fact that you are peering between your wingtips into an appalling, mile-deep gorge and imagine that you are standing on the flat floor of an enormous seabed, a vast plain of lava and sedimentary deposits. Take the geologists' word for it that through compression, chemistry, and time, all the loose, drifting dirt around you gets turned into solid rock. Now, in the mind's eye, move the earth's crust around underneath all this rock so that eventually it breaks apart in fault blocks and gets thrust up into high, snow-capped mountains.

Imagine more heat and pressure, and turn the bottom-most sedimentary deposits underneath the mountains into a metamorphic rock called schist; then bring on the eroding agents of wind, ice, and rain. Do this for 300 to 400 million years, until everything but that base of schist is worn away and you are back standing once more on the flat plain.

Now repeat the above—except for the next phase of geologic history we need to make the sedimentary deposits over 2.5 miles deep and begin to sprinkle in some unicellular algae, nautiloids, ammonoids, trilobites, and little bony fish. And, of course, as the seas advance and retreat, advance and retreat, they'll lay down a lot of non-marine deposits like river silt and blowing sand. The earth's crust will move again, fracturing and tilting the surface, creating another range of fault block mountains, erode them away. No hurry about all this. In fact, there are about 570 million years to get the job done.

Perhaps it's enough just to conserve the essential information and forget trying to imagine the scene metaphorically. We need only remind ourselves, when we stand at the top peering down, that we are looking at what are, in effect, geological tree rings from the two oldest geological eras, ranging from 4.5 billion to 225 million years ago.

At the very bottom of the pile, in the deep gloom of the Inner Gorge where I sit and ponder all this, are the Vishnu schist and Zoroaster granite of the earliest period in geologic time. About two billion years ago, sand, silt, mud, and clay began accumulating at the bottom of a Precambrian sea, eventually attaining a depth of nearly 10 miles. When the region underwent a period of tremendous mountain-building upheaval about 1.7 billion years ago, these sedimentary layers of shale, limestone, and sandstone were infused by liquid magma from beneath the earth's crust, and under great heat and pressure both igneous and sedimentary rocks metamorphosed into the schist and gneiss we now find exposed. And though it may be hard to imagine, there was a range of alpine peaks where there is now a "dreadful abyss." These black, cheerless, metamorphic rocks of Upper, Middle, and Lower Granite Gorge are, in fact, the remnant roots of a once-towering mountain range.

September 29

This morning, after about an hour on the river, we drift down on Vulcan's Anvil, a towering lava rock just off the right bank. There is a niche in it on the downstream side where reverent boatmen have placed small offerings to the god of fire, that angry deity responsible for the ordeal they are about to endure. We are gently amused to observe that it contains a condom, a half-smoked joint, and a wilted flower from the sacred datura plant. Sex and drugs. What more could Vulcan want?

Gradually, it occurs to me that coming upon this icon in the midst of this expansive pond of slack water means . . . major constriction ahead . . . means river backing up before . . . before . . . Holy Mother of Christ . . . *Lava Falls*. Of all the "big drops" in the American West, Lava is the biggest—an abrupt, 37-foot plunge through your nastiest bad dream, a nightmare mélange of basaltic boulders, titanic waves, reversals, holes, a mere eight-second ride, but hang on and hope you got it right going in because there's not a God-damn thing you can do about it if you didn't, except pray and swim.

The most experienced boatmen don't just get butter-flies; they get ravens, condors, griffins. My stomach feels as if I've eaten live rodents. What is left of my mind reels, and I begin to consider whether I should ask that my body (assuming it is ever recovered) be taken up a side canyon and left for the scavengers, or whether I really do have responsibilities back in the tangible world that demand my immediate attention and, hence, evacuation by helicopter to the Toroweap Ranger Station. The Howlands and William Dunn abandoned Powell at Separation Rapid; wherein is it writ I have to stay with my chums at Lava?

Let us not indulge in hyperbole. Let us simply say that Lava Falls is my personal reality check; the more I study it the less appealing the Grand Canyon becomes and the more I eschew the notion that it might be terribly romantic to just disappear into the southwestern landscape like Everett Ruess. The "snug safety" of former ruts begins to look better and better.

But I survive Lava Falls, even run it with a certain

je ne sais quoi. Elan comes to mind. Jim Slade con-gratulates me on the beach at the bottom by observ-ing dryly, "Well, Doctor, you just passed your Ph.D. exam," and I float toward Diamond Creek feeling rather pleased with myself after all . . . smug comes to mind . . . a warm spot growing again in my heart for this muddy old stream. It has carried me into the Great Unknown, and through.

As a matter of fact, I find this canyon the most sub-lime piece of unreal estate on earth and can't imagine why I'd ever want to be anywhere else. It is a common experience among river travelers, I'm told, to lose (gen-erally about the third day out) their essential connec-tion to the "normal" world—what time it is, what day, even what month. Obligations back home are forgot-ten, along with heretofore critical issues like who won the World Series, when will the stock market rebound, have the Bushies started World War III, is my divorce final? The house may be on fire and the children alone, but it's all a matter of supreme indifference. No phone, no fax, no e-mail, no radio, TV, *Wall Street Journal*. Since there is nothing to know about, there is nothing to worry about. It's a seductive state of mind. I begin to contemplate ways to make it permanent.

OCTOBER 2

After Lava we spend a day relaxing, catching up on our journals, reading. Lynn and I fight our way through a thicket of tamarisk and hike up a small side canyon where we encounter . . . more tamarisks. It would be gratifying to announce that along the Colorado River and its tributary streams the dominant "indicator" plants are still the seep willow, coyote willow, desert broom, and Fremont cottonwood, but sadly that is not true. The dominant plant is this noxious foreign weed called tamarisk (actually a bony tree impersonating an anorexic shrub) that was imported from Asia into the Southwest in the mid-1800s as a form of erosion control. It has metastasized up every river in the American West, choking out native vegetation, destroying the indigenous habitat, taking over beaches, and consuming vast quantities of water with its interminable root system. One must not be deceived by what some of the guidebooks refer to as the "feathery grace" or the "delicate perfume" of the tamarisk. It is a biological nightmare, an abomination, the Bela Lugosi of herbal life.

Fortunately, tamarisk confines itself to the banks of drainage systems and hasn't totally crowded out indigenous species just back from the river. One still encounters varieties of "characteristic" Lower Sonoran vegetation—like mesquite, catclaw acacia, creosote bush, four-winged saltbush, and barrel and beaver-tail cactus. Of these natives we pay highest tribute to the creosote bush, a truly sociopathic plant that exudes a toxin into the ground around itself to ward off competition for the infrequent rains that sustain it. Given the chance, it will even poison its own off-spring. Small wonder that specimens of the creosote shrub have been determined to be the oldest living things on earth, even older than the bristlecone pine.

Over the past few days the canyon walls have been receding, the rims dropping lower, the river running more quietly. The evening sky displays an even brighter bowl of stars. Time dissembles. It is a temporal condition born of our illusion of isolation, I think—which is quite astonishing when one considers the reality. Last year, nearly five million people visited the two rims of the Grand Canyon, and over a million actually ventured down into it either by mule or shank's mare. Twenty thousand people floated the river on rafting trips. Some 800,000 viewed the park from one of the nonstop "scenic air tours" (primarily helicopter over-flights) that rattle the canyon's solitude more than 70 percent of the time. In fact, Grand Canyon Airport is the busiest airport in Arizona, except for Phoenix, and daytime conversations in Tusayan, near the South Rim, are conducted in short bursts between the racket of helicopter departures and landings.

During the summer months, up to six thousand cars a day can be observed endlessly circling Grand Canyon Village, fighting for fewer than twenty-five hundred parking spaces, and over one hundred bus tours a day, on average, roll through the park. The *Colorado Plateau Advocate*, a publication of the conservation organization, the Grand Canyon Trust, observed that during July 1993 (the busiest month on record at the park), "more than 231,000 vehicles carried 800,000 people into the park. Another 14,000 visitors arrived by train, and approximately 30,000 people roared over the canyon in 10,000 separate air tours."

As far back as 1978, then Superintendent Merle Stitt acknowledged that the government mandate to manage the parks "by such a means as will leave them unimpaired for the enjoyment of future generations" was an unattainable goal at Grand Canyon. He simply did not have the staff or the budget to do it. And in 1978 Stitt was only trying to accommodate three million visitors.

So . . . we are not alone. And shortly our membership in the human race will be manifestly reinstated. At the Diamond Creek take-out, we'll go about the sad but satisfying job of deflating, derigging, decompressing, deconstructing (and soon, one hopes, deodorizing). We'll head up the long, dusty trail to Peach Springs and the highway to Seligman. Where we'll hit the interstate and a time warp that will take us into Flagstaff, with its bars and restaurants and motels, shopping malls, health food stores, organic veggies, decaf double lattes, and Szechuan stir-fries. Where a Southern Pacific freight train rumbles through town every fifteen or twenty minutes. Where we'll quickly understand that what we most want is to proceed right back up to Lee's Ferry, have dinner at Marble Canyon Lodge, and launch our boats once again into the madcap unknown. But we'll settle for a Holiday Inn and a hot shower.

DEEP ECOLOGY

THE STUDENTS in the back of the van think I'm nuts. For the past three hours I've been talking about nothing but the Navajo taco I'm going to consume at the Golden Sands in Kayenta, and to them a taco is a Mexican taco—a tortilla with meat and cheese and some shredded lettuce in it, maybe some avocado. They have no curiosity about the Navajo taco, no gustatory memory of fry bread and beans. In point of fact, they have very little curiosity about anything foreign to their sub-adult palates, their taste buds having been destroyed by tofu, alfalfa sprouts, yogurt, and herbal tea.

These students seem to evidence very little curiosity about anything at all. From Monterey Bay to Flagstaff, a distance of nearly 900 miles, they have lain supine on the floor of the van, sleeping, rummaging through my Conway Twitty, Merle Haggard, Charlie Pride, and Melba Montgomery tapes (complaining), and occasionally dipping into private stocks of seeds and nuts. One girl has eaten nothing for two days but garlic-flavored popcorn. The Mojave, beyond which few of them have ever been, holds no interest.

One or two bestir themselves to look at the Colorado River when we cross it, largely because they have read (or are supposed to have read) portions of Powell's *Exploration of the Colorado River and Its Canyons* and because they've been lectured on the fact that this river is the major artery of the entire western drainage between the Sierra Nevada and the Rocky Mountains—a drainage, they dimly understand, that we will be floating when we put onto the San Juan river near Bluff, Utah.

"Looks dirty," the popcorn eater remarks.

The San Francisco Peaks above Flagstaff inspire a yawn, as do the Painted Desert and the Little Colorado. Black Mesa, however, produces a communal outpouring of invective directed at Peabody Coal and Uncle Tomahawk Native Americans who conspire in the rape of Mother Earth. These students are, after all, majors in environmental studies, and although they take little interest in the actual environment, they are not short on opinions about its defilers.

We reach Kayenta about five o'clock, and I see Bud's truck parked in front of the Golden Sands, its trailer load of boats, rowing frames, oars, coolers, and miscellaneous gear in marked contrast to the more labor-oriented contents of the local Indian pickups. A patchwork of bondo and rust, it has once again earned its reputation as "the puker": I notice most of its human cargo sprawled in various angles of repose around the parking lot, exhaling carbon monoxide and trying to regain their stomachs.

My group, as usual, begins what is for them a laborious process of democratic resolution—will they eat, or will they wait in the van?—but their need to hold a town meeting pursuant to action (any action) soon defeats me, and I head for the restaurant alone. One of the remarkable things about this outfit, I have come to understand, is that no one will commit to anything unless everyone commits.

Bud is inside concentrating on the purpose of our pit stop—the Navajo taco. The *large* Navajo taco. A massive, mammoth, monstrous, Falstaffian, Brobdingnagian, gargantuan, Cyclopean fatty of a taco served up on a plate the size of a turkey platter and weighing about 25 pounds. An acre of fry bread, a bushel of beans, a furlong of cheese, a firkin of lettuce ... God knows what else. I can't see the top of his head behind the escarpment of his victuals, but I can hear heavy breathing and sybaritic moans.

The student consensus, apparently, is that it is too early for chow, and most have elected to cool their heels in the parking lot with the refugees from the puker. Four of the more adventurous ladies wander into the restaurant, through the melee of Kayenta Navajos, and sit silently at one of the few empty tables. I observe them shake their collective head when they are handed menus. Nothing to eat, thank you. We will have four glasses of water. The waitress regards them for a moment without emotion. "You don't order, you don't sit," she says. They sigh, look put-upon, rise. As they make their way to the front and are about to exit, a young Indian with a walking cast on his left leg comes through the door. The popcorn eater nearly runs him down. "Excuse me," he says, lurching back. No response. Not only does she ignore his courtesy, she doesn't even notice him. He is vapor, wind, a figment of her imagination; he has to flatten himself against the wall to avoid getting knocked on his behind. "Excuse me again," he mutters.

Bud watches this cultural interface, slowly masticating the last wad of his fry bread and beans. "Maybe we better round up the wagons, Kemosabe," he says. "One of these dog soldiers is likely to give offense."

"You round up the wagons," I tell him. "I'll catch you in Mexican Hat. Because I'm gonna eat my taco, I don't care what."

Bud inspects the toothpicks in a shot glass on the table. "That's exactly what Custer said to Reno down there on the Little Bighorn," he says. "So maybe you'll catch me in Mexican Hat. Then again, maybe not."

Back on the road. Great thunderheads over in the direction of the San Juan Mountains, and rain squalls streaking the sky around Mesa Verde. Or maybe it's just fallout from the Four Corners power plant. All around us the de Chelly sandstone buttes of Monument Valley are ablaze in the late afternoon sun. One of my wards is moved to crawl up off the floor and ask what makes them red. "Iron," I say. He wants to know how they got here in the first place. "Erosion," I tell him. Yesterday I might have been up to a more expansive discourse, might have bored him with the little information I possess about the intrusion of ancient seas and the deposition of sedimentary beds; about coral reefs and biothermal banks; about upwarping, downwarping, slumping; about river cutting, wind, spheroidal weathering, oxidization—nifty stuff like that. But this evening I just want to drive across the "rez" with my own head for company. I'm beginning to wonder if joining this expedition was such a good idea after all.

Who knows? It would seem that taking students who study the environment out of the classroom and into the "field" (I should say "down the river") ought to be an act of true pedagogical devotion. Either that or the greatest academic scam ever conceived—rafting on the taxpayer's dime, so to speak. Of course those of us who are merely serving here as "guides" don't have to concern ourselves with such hairsplitting—don't have to do anything, in fact, but drive the trucks and row the boats. Professor Pshaw, who should be waiting for us at the put-in at Sand Island, has done all the planning and outfitting. He's the one who will give the lectures and lead the hikes. He's the one who has to worry about one of these narcoleptics doing a head-plant off a cliff (ground balls, we used to call them in search and rescue). Drownings, broken bones, hyperthermia, PMS, snakebite, scorpions, fire ants, fire pants, impregnations, drug abuse—all his responsibility. The rest of us just have to keep this torpid, temperamental, hormonal mass moving in more or less the same direction—a chore, Bud has observed, rather like trying to direct a centipede through a maze, one leg at a time.

Dusk is upon us as we loop over the north end of the Raplee anticline a few miles southeast of Bluff. The great Comb Ridge monocline lies just in front of us, 80 miles of abrupt cliff face that marks the eastern boundary of the geological formation across which we have been bouncing for the past hour. The Monument Upwarp, as it is called, is a kind of natural superdome, 35 miles wide and 100 miles long, between the Colorado River on the west and the Paradox Salt and Blanding Basins on the east. Its northern definition begins approximately at the confluence of the Green and Colorado Rivers in Canyonlands National Park, Utah, and its southernmost extension is the Golden Sands Restaurant in Kayenta, Arizona. Well . . . near there anyway.

Bisecting this wasteland of wrinkled rock and treacherous little thorny plants is the canyon of the San Juan River (canyons, actually), cutting across the Grand Gulch Plateau, down through the Permian to the Pennsylvanian, exposing on its way a host of stratigraphic terms that basically describe time deposits of limestone, sandstone, shale, siltstone, marine organisms, layers of this and that—a kind of geological Navajo taco. I can never remember half the ingredients, much less the order of their spread. I can never remember whether the Cedar Mesa formation is on top of the Halgaito, or the Halgaito on top of the Hermosa. Or all of the above. Is it Moenkopi shale that caps the de Chelly sandstone, or Sinarump? And things like the simple distinction between, say, a syncline and an anticline just flat out elude me. I have to conjure the letter *A* (for "anti") in my mind's eye and translate it to my finger, draw a diagram in the dashboard dust. The slopes meet at the top in a picture worth a thousand words.

But such details are of limited importance. I want you to sit up back there, you louts, and take notice. What is before you in this failing light is not scientific nomenclature; it is the most staggering image of cliffs, washes, canyons, buttes, mesas, towers, cathedrals, walls, potholes, draws, swells, folds, pockets, cones, spires, needles, and labyrinths you're ever likely to see. Attention must be paid.

The San Juan River flows quietly between its banks at Sand Island, gurgling occasionally in the darkness when the subsurface current decides to boil up for a look-see. No telling what the river gods are doing out there. Once on the Rogue River in Oregon, I had one of those random boils suck down the rear tube of my raft before it decided to let go. But not on the San Juan. The San Juan is a gentle float without serious hydraulics. No rapids worthy of notice.

Magnificent, towering walls, sandy beaches, hot sun and smooth rock, cottonwoods, the invader tamarisk, canyon wrens. The San Juan is distinctly a mellow experience ...

"Except that we've got a problem," Bud says, coming

out of the campfire light where he has been overseeing the preparation of supper and into the riverbank darkness where I have been hiding. He sits on a pile of life jackets unloaded earlier and rolls a smoke. "We've got five vegetarians on board. They're caucusing right now about what they delicately describe as the 'nutritional inadequacy' of our commissary. They want to go into Blanding and buy tofu."

"Tofu! Blanding is thirty miles . . . and they're not going to find tofu in Blanding. Tofu?"

"I told them. They say they can't go five days without an acceptable source of protein."

"We've got all kinds of protein. Eggs, cheese, nuts, tuna fish. What do they usually eat?"

"Tofu."

"Jesus."

We walk down to confront the congress gathered just outside the kitchen area where the other guides, Lynn and Don, are grilling burgers. Bud explains that going to Blanding is out of the question and points to all the protein goodies we already have in the dry boxes—eggs, cheese, peanut butter, beans. "Those of you who don't eat meat can load up on beans tonight," he says. "We've got a huge pot going; you can eat beanburgers, salad, fruit, cookies."

"We're not in the midst of civilization, folks," I add. "We'll just have to make do."

A girl named Chanterelle steps forward and eyes me malevolently. She is one of the smokers on the trip— she and a frail asthmatic kid everyone calls "Fuckin' A

Fred," though the connection between this sobriquet and its object is opaque, to say the least. Chanterelle, however, looks a lot like the mushroom she is named after—flat-headed and short-necked, shoulders like a nose tackle, no waist, no hips, no glutes. All stem from the armpits down.

"What kind of beans are they," she says, letting me know by her inflection that I am about to learn something.

"I don't know . . . beans are beans."

Chanterelle produces the empty number 10 can and holds it up for inspection. "Have you read the label? Ranch beans!" she intones. "Cooked in pork by-products."

Great God. Skewered by a food Nazi. We can only shrug and walk away, hoping everyone recognizes a Mexican standoff. "Bad'ges? We don't need no stinking bad'ges." Anyway, this is Professor Pshaw's problem, not ours. We just drive the trucks and row the boats. "Where is Pshaw," I ask Bud, "in our hour of travail?"

"At the Recapture Lodge in Bluff."

"Doing what?"

"Lodging. He said he's slept on the ground before. He said he'll be here in the morning to help load the boats."

The moon has come out. Same old moon, I imagine, that the Anasazi admired when they lived in these canyons as far back as two thousand years ago. They did all right on beans. A complete protein, the bean, when mixed with a little corn and squash. Freed the ancient enemies from all that hunting and gathering and hitchhiking into Blanding for tofu. Gave them time to settle down a bit, take up the arts. In fact, tomorrow we'll stop a few miles downriver to look at a whole wall of their art (petroglyphs carved into the Navajo sandstone at the mouth of Butler Wash) and to pick through a field of their pottery shards scattered around the base of the cliffs. I must remember to instruct the children to put everything back where they find it. It's a no-no to steal samples.

There were at least three separate periods of Anasazi occupation in the San Juan drainage—roughly AD 200–400, 650–700, and 1050–1275. And then rather suddenly they left. Why they left is a matter of some speculation, but climate was probably the major factor. Tree ring counts and pollen studies in a number of granaries show a decrease in rainfall that reached serious proportions during the last quarter of the thirteenth century. Twenty-five years of drought coupled with a long-term population increase undoubtedly spawned a host of ancillary problems—overuse of depleted land, over-irrigation, increased erosion, reduction of game animals, reduced nutrition, and a resultant susceptibility to disease—all that plus a growing paranoia that the guy in the next gulch over might be plotting a raid on the food cache. Whatever the specifics, by about 1300 they were gone. And except for Fathers Escalante and Dominguez in 1775, and a few trappers in the 1840s, nobody came here again until the Corps of Topographical Engineers under Captain J. N. Macomb in September 1859. Macomb was not overly impressed. "I cannot conceive of a more worthless and impracticable region," he said.

We commence our float around mid-morning. Lynn and Bud will row the two bright yellow Domars, Don the Avon Pro, I the old Achilles. We will each take three passengers except Don, who gets the bonus extra— Professor Pshaw. The well-rested doctor makes his appearance at the last moment (looking badly hungover, to tell the truth), but manages a brief discourse on the genesis of "desert varnish" before dropping a wet beach towel over his freckled pate and retiring to the shade of a cottonwood tree. The guides marshal the centipede into the boats, give instructions about life jackets and sunstroke, and shove off.

Rendered mute by the feeling of release that comes with departure (all hype at last hypostasized), we slide quietly past low banks of graveled terrace and lean back at the oars to gaze on the flat-topped mesas encircling the river valley around Bluff. Boat bottoms scrape occasionally against submerged spurs of midchannel sandbars. The current wanders. Seven miles to the west, the river will slice through the Comb Ridge monocline and speed up its twisted descent across the Monument Upwarp toward the Colorado River, but here it is a slow meander, a good place to let the raft drift through lazy 360s. Work on the tan.

The students have decided to give themselves nick-names. As I float down on the Achilles I hear passengers shouting back and forth across the water.

"Hey 'Shrooms, you got my number 8?"

The unmistakable conformation previously known as Chanterelle rises from Bud's thwart tube. "What? Say again, Beaver?"

"My sunscreen. You got it?"

"Gave it to Warbler."

"Yo, Warbles, 'Shrooms says you got my sunscreen. I need it, man; I'm turning the color of a crawdad."

"No way, Beav. 'Kin' A had it at the put-in."

The boy called Beaver turns to the apparition next to him, something shrouded from hood to hoof in a white nylon rain poncho. Must be 200 degrees in there. "Hey, dude, you got my number 8?"

"Fuckin' A," says the wraith, poking it out through the armhole of his tent.

Warbler? Beaver? 'Shrooms? The current catches my boat and carries me past the Domar. I look for expression beneath the baseball caps that Bud and Lynn wear low on their foreheads, see only the glint of river and sun in the lenses of their mirrored shades.

The great cliff of petroglyphs at Butler Wash is a howling success, less for the mystery of its symbolic representation than for the manner in which it has been defaced by modern scriveners recording names, dates, sweethearts, and hometowns in the soft surface of its ancient rock. There is also a spray-painted message across the length of the wall, "River Runners

Go Home." Like the strip mine at Black Mesa, Butler Wash inspires outrage, as well as loudly expressed
opinions about the disposition, percipience, and cultivation (not to mention lineage and pedigree) of the
Caucasian geeks ("probably Mormons, probably from
Moab") who carved these pitiful forgeries into the face
of time. It seems pointless to brand oneself a racist by
suggesting that the culprits are as likely young Navajo
as Moabite whites—young Navajo who despise river
runners (most of whom are white) and who have no
reverence for the Anasazi either. The Anasazi are not
the ancestors of the Navajo. Only the word is Navajo.
Meaning "ancient enemies."

Interest in pot shards terminates with the call for
lunch. The discussion around the tuna salad (PB&J
for nonusers) turns to the Grateful Dead. Chanterelle
and Warbler continue to demonstrate their contempt
for the bill of fare and boycott the table, choosing
instead to sit in a thicket of tamarisk and smoke per-
fumed cigarettes. I take my sandwich and walk up the
talus slope below the wall to eat in the company of
one of my favorite rock art characters—a little trapezoidal man (or woman) with a little trapezoidal head
and little stick arms, legs, and fingers. Inside his trapezoidal head stands a still littler trapezoidal man (or
woman)—with all the appropriate appendages. What
is the meaning of this? What is that second fellow
doing up there in that head? A mystery. Somebody
once told me that the Navajo word for "soul," directly
translated, means "the one who stands within me," or

"the one who guides me from within." Is that what we have here? Some Anasazi scratched a picture of his soul into the oxidized face of this ancient seabed? Totally far out! Intense . . . as the young persons say.

We make our first camp at mile 10 below the Mule Ear diatreme. Dr. Pshaw seems to have recovered from whatever ailed him at the put-in (the nap he took at mile 6 while the rest of us hiked to the cliff dwellings must have helped), and he steals a text from *San Juan Canyons: A River Runner's Guide and Natural History of San Juan River Canyons* by Donald L. Baars and Gene Stevenson. Baars tells us (in Pshaw's voice) that the Mule Ear diatreme "is a kimberlite-bearing diatreme and contains a great variety of crystalline rocks from the Precambrian basement complex ranging from coarse grained granite to gneiss to serpentinized talc-chlorite schist." Pay attention, scholars. There will be a quiz. "The presence of eclogite with dunite, pyroxenite, peridotite, and large blocks of kimberlite suggest that the separation of the gas phase took place at considerable depth, possibly near the crust-mantle boundary." Now . . . if there are no questions? . . . we are going to hike to the top of this volcanic vent to see if we can find any small red garnets lying around in its seventeen-hundred-million-year-old rubble. Look in the ant hills. The fire ant hills. Do not provoke the ants—or they will make small red garnets on you.

As the students prepare for departure, the vegetarian delegation approaches the kitchen. A swarthy, dark-haired boy now known as "TV" (for turkey vulture) asks Bud what they can expect in the way of sustenance when they get back from the "dithyramb."

"Roadkill," Bud says.

"Sir?"

"An old river tradition. First night out we always have roadkill. We got lucky at the junction of 89 and 160, so tonight you get a choice—snake tire-tire or treaded veal cutlets."

As it turns out, this is the last discussion we will have about ingestion.

Expulsion, on the other hand, is the subject of the evening performance—Bud and Don demonstrating the proper use of the WC (wilderness crapper), also referred to as the port-a-pot or the groover. Regardless of its designation, its construction is always the same—a steel rocket can, double lined with plastic garbage bags, and a removable toilet seat. Instruction is needed because (1) some people are shy, embarrassed, and revolted; (2) a restrained use of Clorox and lime disinfectants is required ("You are not baking bread," Bud always says. "Do not flour the pan."); and (3) the groover accepts only solids. "Do not pee in it," Bud intones. "Pee in the river."

There would be little point in bringing this subject up had it not provoked our band of merry travelers into yet another attitudinal outburst. An endorsement of child molestation could not have been more enthusiastically received. The chorus of boos verily echoed off the adjacent cliffs—pee in the river . . . *boo*. Bud remained calm. "Pee in the water, or in the damp sand beside the water, not in the groover. Everything we bring with us, we take with us—except liquid."

"Boo. Polluter."

"When we get into the Goosenecks this practice is particularly important because the beaches are small and narrow, and the plant life is fragile. About 6,000 people a year float this canyon, and if every one uses the area just around camp as a latrine, it becomes a very smelly affair indeed."

"Don't listen to him. Boo."

"You better listen to me," Bud says. "Because if I find liquid in the groover, you'll find it in your morning coffee."

When the moon comes up, I walk down to the beach to check bowline knots and make sure everything is secure. Night winds have been known to whisk unfastened tarps, life jackets, clothing, even boats themselves into the river, and I have encountered some screamers in this canyon—night and day. At the Clay Hills take-out I once watched an unloaded 16-foot Avon picked up by a tremendous gust and blown like a leaf (or a barn door) for 200 yards across the parking lot, flattening a half-dozen rafters along the way, and taking out a loaded picnic table set up by a commercial outfitter for the delectation of his hungry customers. And more than once I have spent an irritable night sweltering in the bottom of my sleeping bag trying to avoid being sandblasted by a San Juan sirocco.

Don and Lynn are battening down the kitchen and discussing itinerary with Professor Pshaw—arguing a long river day tomorrow because we want to get at least to the entrance to the Goosenecks below Mexican Hat to camp. The students have retired to a tent ghetto they established before dinner, a circular arrangement with the opening of each hovel facing in toward the center. Bud is on his raft drinking a beer and trying, by flashlight, to unthread a nut off a bent thole pin. I sit on one of his tubes and dangle my feet in the water, looking up at the moonlit spine of the Lime Ridge anticline. "Why do you suppose they're all sleeping in a circle?"

Bud turns off the flashlight and stares out across the river—the gray-green, greasy Limpopo. "So that they can maintain constant audio/visual contact with each other," he says. "Part of the wilderness experience." He yawns, and tosses the thole pin in his repair kit. "You know, there's something about these folks that really gets to me."

"I can't imagine what."

"They've got an attitude about everything. They don't know anything, but they've read Barry Lopez and Edward Abbey and they think that's all the information they need to get opinionated and self-righteous about the 'out of doors.'"

"Sounds to me like you been quarreling with the clients again."

"Not quarreling, trying to instruct. I told TV to take the soap and wash before doing kitchen duty and he said he didn't use our kind of soap because it was made from animal fat."

"I hope you threw him in the drink."

"Actually, I just threw him out of the kitchen."

Day two, and we need to make some river miles if we are going to get a decent place to camp tonight. Lieutenant Joseph Christmas Ives, who had about the same opinion of slickrock topography as his predecessor, Captain Macomb, commented on the paucity of decent places on the Colorado Plateau in his *Report upon the Colorado River of the West* (1861). The area, he said, "is, of course, altogether valueless. It can be approached only from the south, and after entering it there is nothing to do but to leave. Ours has been the first, and will doubtless be the last, party of whites to visit this profitless locality." Wouldn't it be nice if that were so. Unfortunately, there are a good many parties in this profitless locality right now, all hiking in the same side canyons, all vying for the best ground at night, all scowling at one another as they pass on the river. It used to be considered poor form to be obviously trying to out-row another trip to the primo campsite just around the bend, but now it is the norm. The BLM limits the number of groups that can launch on any given day through a lottery permit system, but the river still gets maximum usage and the old camaraderie that once defined an encounter between hominoids in the wilderness is a thing of the past. "Howdy" is an anachronism in the wilderness. Proprietary resentment is the fashion. "Who are those sonsabitches?"

So we log miles, while the students sprawl on the raft

tubes and doze. The sun beats down, but there is a nice breeze that cools the sweat—and dupes the dreamer as he slowly dehydrates. Repeated warnings have been given about hyperthermia, but they have been received by our young hotspurs with the same attention paid all such instruction. "You ought to drink some water," I tell the people in my boat. "The desert is deceptive. Even when you think you're cool you're losing a lot of moisture." Patricia Clotworthy ("Cow Patty") raises her head and scowls, thereby exhausting her range of expression. The others sleep on.

Occasionally Professor Pshaw, riding in the lead boat, requests that we stop—twice for people to go into the bushes, once to inspect evidence of a fossil oil field called the Ismay algal biotherm ("see the leached oolites? That's the top of the Desert Creek cycle."), once at Mexican Hat rock. There is a short trail to the base of the hat that the guides and four or five students run up. The rest of the party huddles in whatever shade is available along the bank. We make a fruit salad for lunch during this interlude, and when we're done Bud throws the residue in the river. Chanterelle regards him as if he just shat on the Eucharist wafers. "The catfish and suckers will eat it," he says. "It's biodegradable."

"I wish you wouldn't do that anymore," she says, exhaling a cloud of cigarette smoke. "It's very offensive."

Twice during this layover I catch "the Beav" doing his business up behind the small sandy area where we have stopped. Twice I ask him not to do it again. He responds by briefly contemplating the rim of the canyon and walking away.

Our hopes for the good camp at mile 29 are dashed when we pull up on the beach under Mexican Hat Bridge to refill our 5-gallon water jugs. The highway crosses the river here, and a short, rocky road down from the trading post above has been blasted out of the shale, making this a relatively quick and easy stop. It is even possible, with a little nerve, to drive down to the water's edge—indeed, two pickup trucks are parked at the far end now, two families of Navajo occupying an area next to the cottonwood trees. The kids frolic in the water, the women sit stolidly in the cab, and the adult males hunch under the tailgate and stare out at the kids from under baseball caps that advertise the main reason they have come off the reservation. Budweiser. Coors. They are drunk. And they are not friendly. And when their kids start climbing on our rafts, they call them sharply off. "Get away from those people. They got VD."

Under the circumstances it is, of course, impossible to keep our kids away from their (metaphoric) raft. It now becomes essential to demonstrate that the white brothers and sisters understand. The white brothers and sisters want to confirm the validity of the insult and to assume full responsibility for the fact that the red brothers are wasted on a Tuesday afternoon, sitting in the dirt under a pickup truck with a six-pack and a brown bag and looking like they can't decide whether to puke, pass out, or go completely berserk. The white brothers and sisters would like to share the pain, the anguish, the all-consuming and uncompromising rage.

The Navajo, however, don't want to share anything. Including the same air. Custer's offspring are a familiar pain in the ass and can be either sidestepped, ignored, or swatted like a horsefly, but they're too dumb to be insulted. Insults merely inspire them to higher levels of obsequiousness. And now that they've got their quarry trapped at the end of the beach, they can't even be sidestepped or ignored. Perhaps they can be run over. If they won't leave us alone, we'll leave them alone. Excuse me, again.

As Bud and I are returning from the trading post with the water jugs, we can see there's trouble down below. Evidently the Navajo have tried to depart by backing along the beach (no doubt at full throttle), and have been thwarted by Demon Rum and their own tempers. One of the pickups is half on its side, right rear wheel in the river and the other madly spinning a continuous plume of sand into the boats; the second pickup has a chain attached to the first and is creating its own grit storm as its tires scream and smoke on the loose shale of the ramp. A pale clutch of white brothers and sisters cringe behind the cottonwoods, confused, wanting to help but dimly aware, at last, that help is not wanted. The Indians are cursing as they try to avoid being decapitated by the shrapnel fired from under the tow, while at the same time they strain to keep the towee from completely collapsing on its side. The truck drivers (now both women) seem to have but one purpose in mind—to mash the hammer to the floor until things either come unstuck or the engines blow. Bud sets his jerry can on the ground. "How do you feel about let's go get an Eskimo Pie?" he says. "Let things sort of work themselves out."

Except for the loss of the campsite caused by this minor delay, we are not otherwise inconvenienced by our skirmish. The scene under the bridge seems to chasten the California cosmologists for a time, though there is heated discussion around the postprandial fire to the effect that the Native American predicament, as evidenced by the afternoon's events, is directly attributable to long-standing Bureau of Indian Affairs (BIA) paternalism and a resultant confusion on the part of the Indian as to his status. This leads TV to a definition of Indian status. The Indian is the "first ecologist," he says. There is a collective murmur of affirmation. No one questions the absence of stitching in his segue. One brave lad named Wyckham Snavely offers the unpopular opinion that Indians have been as guilty of clear-cutting, overgrazing, strip-mining, wasteful hunting and fishing practices, and resort development as anybody else, but his argument is spiked by Deadhead Darleen before it gets off the ground. "You're a racist, Wyckham." Touché! Before long the circle around the fire tightens and Wyckham, finding himself squeezed, goes off to bed. Thus we deal with the Pyrrhonist.

Day three, and we float through the Goosenecks,

that curvilinear, meandering section of the San Juan that begins at the Mendenhall Loop and continues for 30 miles to John's Canyon, where things begin to run a little straighter, northwesterly/westerly, for another 23 miles to the take-out at Clay Hills crossing. The canyon is deepest here, 1,235 feet deep at the foot of the Honaker Trail, where we spend an afternoon climbing the 2.5 miles of switchbacks built in 1904 by a gold prospector named Henry Honaker. Some of us, anyway. Cow Patty and Deadhead Darleen are suffering from early symptoms of hyperthermia and have to be attended to by Lynn (male concern simply elicits tears and irritability). Others take one look at that monstrous, near vertical wall of three-hundred-million-year-old sandstone, limestone, and shale and decide to wash their hair. Wilderness exploration goes on the shelf when it comes to hair.

Too bad. Because the view from the top is unparalleled—the Abajos and the Henry Mountains rise over 11,000 feet to the north; the snow-capped San Juans rise 13,000 and 14,000 feet to the east. Behind us Monument Valley spreads out in subtle shades of ocher, backdropped by the dark, looming mass of Navajo Mountain—which is backdropped in turn by a towering bank of thunderheads coming in from the west. And to our left, off in the direction of the Colorado River, lies the vast expanse of the Glen Canyon National Recreation Area, 1,200,000 acres, decipherable in the late afternoon light only by shadow and horizon. We do not stand close to one another up here on this stone cap of the earth. We seek private spots for private thoughts, albeit private thoughts with a common base. Alone in the wind and rock, it is perversely comforting to acknowledge, however briefly, one's utter chronometric and horologic insignificance. We don't matter; therefore nothing matters. It takes a great load off. Relieves us of a great freight of pompous responsibility—before we dive back into our crack in the ground and return to the river.

Where our thoughts are less transcendental. Simple matters prevail, like when to eat, where to sleep, where to set up the groover, who are those sonsabitches?—they better not pull in here. On the San Juan there are over three hundred campers stretched out along less than 100 miles. Almost three people per mile. About five to six thousand each season. Outrageous. Given the circumstances, the relatively pristine quality of the riparian corridor is truly amazing. There is a reason for this, which has nothing to do with an environmentally sensitive user public. It has to do with regulations that require the use of fire pans for campfires (and the removal of charcoal); bags or containers for carrying out all bottles, cans, and wet garbage; and the deployment of the portable pot. Why do the custodians of our national playgrounds not provide us with well-spaced, semi-permanent, chemical johns, you ask? On some of the filthiest rivers in this country, they do. The problem is that many folks are called but few are chosen at the specific moment an outhouse floats by. The groover floats with you.

On the morning of our last day, the sky emerges pale lavender above the cliff walls, unbroken and cheerless without sun. The night chill does not dissipate, and the students are slow getting up. I pull on my filthy old Patagonia jacket before crawling out of my bag, and squish down to the kitchen area on flip-flops that feel like two cold pieces of liver on my feet. Bud and Don already have the coffee going, and we stand with steaming mugs, staring stupidly at the river and listening to a canyon wren pipe his clear, descending notes from somewhere in the rocks behind us. Gradually we achieve a functional level of consciousness and start to consider breakfast. Bud takes our big chili pot and a metal spoon and goes off to perform what has become his favorite chore—banging loudly and repeatedly in the center of the tent ghetto in order to rouse the inhabitants in as irritable a frame of mind as possible. Lynn appears, yawning, an "Oh, gosh, you're all up, shame on me" look on her face, grabs a cup of coffee, and disappears with her towel and toothbrush. The camp slowly comes alive. Don and I start cracking eggs.

There is a lot of loud, angry shouting going on over by the tamarisk where three or four campers pitched their tents last night (the great circle of conjugates having apparently been disbanded). Pshaw, as usual, is still in the sack, and the three of us drift over to see what the fuss is all about. Chanterelle is berating Wyckham for having smashed a scorpion with his shoe. He protests that it was crawling on his ground cloth and he didn't want to get stung. "It was out to get me," he says, in a fatal attempt at levity.

"That is totally anthropomorphic," Deadhead tells him. "Totally. Why do you ascribe your own miserable aspirations to that poor insect?"

"Because he was going to sting me," Wyckham says, "and because he's not harmless."

"He?"

"Well, it . . ."

"You're not only racist, Wyckham, you're sexist," Chanterelle says. "You're also an asshole."

It's good to see that some things are normal this morning. "I hate to interrupt," I say, stepping forward, "but breakfast is ready and I want all your gear down on the beach before you eat. We don't have much current left and we're probably going to have to row most of the last 5 or 6 miles, so we need to move it."

Indeed, not far below our camp, at Grand Gulch, the current does give out. Lake Powell backs up the canyon of the San Juan nearly 50 miles from its confluence with the Colorado—that is to say, with the confluence of what used to be the Colorado, now a 200-mile, stagnant, silt-laden reservoir behind Glen Canyon Dam. What puddles up beneath us at this point is just the dead backwater of one of its torpid tentacles.

We pass the mouth of Oljeto Wash, flooded to the base of its first sandstone terrace. When the lake is low and the San Juan still a live river all the way to Clay Hills, this is the finest of camps—broad, sandy, protected—with one of the most enchanting hikes up through high, sweeping walls of Cedar Mesa sandstone. Today Oljeto is a shallow pond, a languid eddy with a Clorox bottle and a plastic plate slowly circumnavigating its perimeter. A half mile below the wash Steer Gulch enters on the right, occupied by two boatloads of nudists who regard our passage with vacant disinterest. After another mile, at Whirlwind Draw, I see Lynn's boat snubbed up to the rocks and Professor Pshaw wildly gesticulating from the shore. Both Bud and I hang up on sandbars trying to pull in, and Pshaw is beside himself when finally, towing ourselves with our bowlines through knee-deep quicksand, we reach the bank. "Hurry up, hurry up, hurry up," he keeps yelling. "We got an emergency. Chanterelle's down."

Down? Drown? What's he saying? We scramble over the rocks to a patch of sand where Chanterelle is indeed down, stretched out on a tarp in the shade with Lynn kneeling beside her, wiping her face with a damp cloth and talking to her in a quiet voice. The other passengers from the first boat hang around on the fringe, looking as if an alien had been discovered in their midst. "What's the problem?" Bud says. We can hear Pshaw back at the river's edge bellowing for Don's boat to "eddy out, eddy out."

"She got stung by a scorpion," Lynn says. "Apparently it crawled into one of the folds of her life jacket and when she finally put it on"—meaning, when I told her I'd put her ashore if she didn't wear it—"it nailed her." Chanterelle moans that she's going to lose consciousness, that her arm hurts, that she's cold, that she's burning up. "The problem is she says she's allergic to insect bites . . . bee stings anyway. But I don't know if this is a reaction to venom or histamine. I guess we could give her some Chlo-Amine and see what happens."

"We could give her a shot of epinephrine," Bud says.

Pshaw is tearing out the remnants of his hair. "What is that, epinephrine? Does that work on scorpions?"

"I don't know. Probably not."

"Well you're supposed to know. You're supposed to be a guide."

"I'll get the medical kit while you folks debate," I say, and jog back to the boat. Wyckham is sitting on a rock holding my bowline, the faintest curvature of a smile gracing his lips. He looks at me without curiosity, but for some reason his disinterest seems manufactured. "Chanterelle got stung by a scorpion," I tell him.

"No kidding," he mutters.

When I get back with the kit, Chanterelle already appears much better and in fact is not going to suffer much more than a sore arm and a case of the woozies. Bud winks at me out of the line of her sight and says, "I hope I remember how to stick a needle in. Only did this once, and that was to an orange."

"You are not, I repeat, not, giving me any shot, buster," Chanterelle says.

"Guess you're feeling better," Bud says.

But we wait a while anyway, rocking on our heels in

the cool shade of the cliff, until our patient's normal bellicosity begins to manifest itself in the suspicion that "that sonofabitch Wyckham" might be responsible for her malaise. Recovery complete, we return to the river.

A faint roll of thunder booms up the canyon from the direction of the Kaparowitz Plateau. The sky has turned dark, and the wind brings the smell of distant rain. Bud crawls across his rowing frame and jams his oars back on the pins as if he's suddenly anxious to get on the river. But then he just sits back limply on his cooler and starts miming the wrist action of a dart thrower. "Actually," he says, "I bet there isn't much difference between a mushroom and an orange. I bet it would have been the most fun I've had on this whole trip."

The rain comes down in sheets at the take-out, making the clay banks where we unload the boats slicker than snot on a doorknob. The California cosmologists look forlornly at the trucks up on the rise as they trudge through knee-deep gumbo with oars, frames, coolers, dry boxes, and Bill's bags. A few revolt and head for high ground, but everybody else, amazingly, bends to the work without complaint. They are, in fact, so cooperative that once the gear is portaged they start to get in the way, and we have to prevail upon Pshaw to load as many of them as will fit into his Volkswagen van and head back to Mexican Hat. From there they will go on to a Holiday Inn in Page, Arizona. Don volunteers to take the rest if we will load his raft and frame in one of the remaining rigs. Our pleasure, we assure him. We'll even clean off the mud.

When they are gone we pack gear into the trailer and Lynn's pickup, whistling while we work, stopping when the storm blows over for a sandwich and a beer. But we find there isn't much left worth eating. No tuna fish, canned chicken, potted ham, lunchmeats, bac-o-bits, pepperoni sticks. There are lots of wilted alfalfa sprouts, limp lettuce, and a few bruised tomatoes. Some avocados the consistency of guacamole. Some fruit. No cookies. We agree it just won't do. The herbivores have cleaned us out. There isn't even a can of ranch beans left to fill an omnivorous stomach. A solution, however, occurs to us. Get this junk loaded and hit the dusty trail—up the anticline, down the syncline, across the butte, over the mesa, through the potholes, washes, draws, swells, pockets, folds, and gullies to the Golden Sands Restaurant, where they offer up a gustatory memory. A great fatty of a gustatory memory. A *large* gustatory memory.

Alone in the van, I put on a Melba Montgomery tape and settle down for the long, slow drive back across the Upwarp to Mexican Hat and south through Monument Valley to Kayenta. The road over Cedar Mesa skirts the prologue to all the side canyons that dump into the San Juan from the north—Grand Gulch, Slickhorn, Johns—drops down into a corner of the Valley of the Gods to the river, and then angles southwest across the Navajo Reservation. A great pileup of afternoon clouds causes the light to slant sideways across the desert, illuminating a pinnacle here and a tower there, bringing a distant butte into sudden, radiant relief. Hail dumps on me as I cross onto the "rez" at the corner of the tribal park, but a mile down the road I am back in bright sunlight.

A Volkswagen van is pulled off to the side near Owl Rock. A dozen people with cameras blazing. Do I recognize this herd? I do. Darleen, Fred, Warbles, TV, Chanterelle, the Beaver, Wyckham, Pshaw. They have, of course, seen me. All waving madly. Better pull over. Might be something wrong with the Volks. Damn, a boatman's work is never done. They crowd around my van, pointing out rock formations, shafts of light, tattered hems of rain clouds, forked lightning on Navajo Mountain. Intense. Far out. Totally nectar. Have I ever seen anything like it? Can I believe it? Is it too much, or what? Outasight. Awesome. We don't want to leave. I nod my head in assent. I agree with everything, with everybody. Cow Patty is actually smiling. Everybody is smiling. "Where was that Navajo taco place we stopped at before?" someone wants to know. Now I'm smiling. Pointing. Just up the road. "We'll follow you," they shout. All in unison. A democratic resolution, perhaps, but right on. Amazing how a wilderness outing can alter one's mind.

MISSOURI BREAKS

A WAY, YOU ROLLING RIVER. But first I'd like to row upstream for a few moments and encapsulate a bit of history by way of introduction to this rather sedate but unforgettable adventure along the upper Missouri. It all begins with a real estate deal cut by the third president of the United States nearly two hundred years ago and announced in a letter to his personal secretary on the fifteenth of July, 1803.

Dear Sir:
. . . last night also we received the treaty from Paris
ceding Louisiana according to the bounds to which
France had a right. price 11 ¼ millions of Dollars,
besides paying certain debts of France to our citizens
which will be from 1, to 4, millions.

Thus did Thomas Jefferson convey to his friend and protégé, Captain Meriwether Lewis, what was perhaps the single most important event in all of American history, the acquisition of 820,000 square miles of the North American continent known as Louisiana. As Bernard DeVoto observed in his introduction to *The Journals of Lewis and Clark*, "There is no aspect of our national life, no part of our social and political structure, and no subsequent event in the main course of our history that it has not affected."

With nearly two hundred years of subsequent events cluttering our historical recall, it is easy to forget exactly why the Louisiana Purchase was (and still is) considered so consequential, but the answer is not extravagantly complicated—it doubled the then existing territory of the United States. What Jefferson's ministers bought from the French was the western half of the Mississippi River drainage basin (except an area claimed by Spain south of the Red River between the one hundredth meridian and the Continental Divide), and while it may have been called "Louisiana" in 1803, today we call it Louisiana, Arkansas, Oklahoma, Missouri, Iowa, Minnesota, Kansas, Nebraska, South Dakota, North Dakota, Wyoming, Montana, Colorado, a tidbit of New Mexico, and a table scrap of Texas. Not bad for 15 million bucks.

All of this new territory, obviously, added an incredible repository of natural resources to the federal treasury, even though Jefferson's critics would continue to dismiss it as useless wilderness, a region of "savages and wild beasts, of deserts, of shifting sands, and whirlwinds of dust, of cactus and prairie dogs." Long after an accurate appraisal based on empirical evidence provided by Messrs. Lewis and Clark, western detractors would go on describing the region as an uninhabitable steppe, hostile, barren, and unfit for cultivation. We need not waste space on the accuracy of their vision.

The Louisiana Purchase enabled (one might say ensured) America's transcontinental expansion from the Atlantic to the Pacific Ocean—an expansion that would, in turn, solidify U.S. claims to Oregon country by putting actual human settlement into the region. When the northwestern territorial dispute between Great Britain and America was finally resolved in 1849, nearly 6,000 immigrants had already arrived in the vicinity of the Willamette Valley alone. And since Oregon in 1849 consisted of everything above the forty-second parallel from the Continental Divide to the Pacific Ocean (Washington, Oregon, Idaho, western Montana and Wyoming, and most of British Columbia), it could be argued that the Louisiana Purchase facilitated the appropriation of an additional 250,000 square miles (the Brits got everything above the forty-ninth parallel, the Americans everything below), bringing the grand total for this particular phase of American expansion to about a million square miles, or one-third of the continental United States.

Understandably, the boundaries of an uncharted, unexplored region equal in size to the existing commonwealth were imprecise, but the northern border of Jefferson's acquisition was generally accepted to be the forty-ninth parallel; its western limit was the Continental Divide—which before the Lewis and Clark expedition was largely a transcendent conception whose reality exceeded anyone's actual experience. The Divide was as much speculation as a hundred other fantasies rumored to be out there west of the Mississippi River—like the shining mountains that rose five miles in a single, unbroken ridge above an endless prairie of grass, or the lost tribes of Israel still wandering about in the Great American Desert, or the Indian clan descended from the Welsh and linguistically indebted to the Celts, or the mountain of salt said to extend for 180 miles across the plains. It was definitely time for a reality check.

Jefferson had long contemplated the inevitable expansion of American settlement beyond the Mississippi, and he had long intended to send an expedition across that territory in spite of the fact that until 1803 it was alternately the possession of France, Spain, and then again France. Indeed, after his election to the presidency in 1801, Jefferson appointed Captain Meriwether Lewis as his private secretary, explaining in his offer to the twenty-seven-year-old army officer, "Your knolege of the Western country, of the army and all it's interests & relations has rendered it desireable for public as well as private purposes that you should be engaged in [this] office."

The private purpose that Jefferson had in mind was Lewis's preparation for leading an expeditionary force up the Missouri River in the hopes that it would prove to be the elusive Northwest Passage. "The object of your mission," his commander instructed, "is to explore the Missouri river, & such principal stream of it, as, by it's course & communication with the

waters of the Pacific Ocean, may offer the most direct
& practicable water communication across this continent, for the purposes of commerce." A trading voyage
to the Pacific Coast was generally a three-year affair,
and the commercial advantage of a water route across
North America, even if it were to include what Jefferson imagined might be a short portage over the "stony
mountains," was paramount in the president's mind.

And so on May 14, 1804, the Corps of Discovery, as
the Lewis and Clark ensemble was called, set forth
on an 8,000-mile journey that would take them, over
a period of twenty-eight months, from the confluence of the Missouri River with the Mississippi to the
mouth of the Columbia River and back. They would
cross the northern plains, all of which, from the Man-
dan villages near present-day Bismarck, North Dakota,
to the Great Falls of the Missouri, was uncharted,
unmapped territory. So too were the Rocky Mountains, which they crossed during the summer of 1805,
and the Snake and Columbia River plains, the Cascades, and the Pacific Coast (near present-day Asto-
ria), where Clark would write in his journal, "Great joy
in camp we are in view of the Ocian, this great Pacific
Octean which we been so long anxious to See."

Looking at the big picture from a contemporary
vantage point, it is easy enough to see that the Lewis
and Clark expedition initiated a rather grand historical process—exploration followed by migration, suc-
ceeded, in turn, by territorial occupation and political
incorporation. It was a modus operandi that would

engage the nation for the next hundred years, indeed, until trans-Mississippi expansion culminated on February 14, 1912, with the proclamation of Arizona as the forty-eighth state of the union.

But what of the immediate consequences of Lewis and Clark's journey? Their exploratory reports would provide Jefferson and the republic with both good news and bad. They would put an end to the myth of a Northwest Passage by demonstrating conclusively, if reluctantly, that no transcontinental water route to the Pacific existed. The dream of easier trading ties with Asia went up in smoke. So did hopes of a fast commercial track between the East and West Coasts.

Their geographical accounts of the upper Missouri, northern Rockies, and Columbia River drainage, however, included detailed notations of the flora and fauna, including the observation that the entire region was swarming with beaver. It was the Corps of Discovery that really launched the American fur trade and brought us such legendary figures as John Jacob Astor, William Ashley, Andrew Henry, Jedediah Smith, Jim Bridger, Tom Fitzpatrick, Joseph Walker, William Sublette, and a thousand other nameless "free" trappers.

Like the Louisiana Purchase itself, the significance of Lewis and Clark's expedition is immeasurable, not entirely because it was the mother of all camping trips, but because it filled in such a huge blank on our provincial map, and it did so with trustworthy details. As DeVoto said, "[Theirs] was the first report on the West, on the United States over the hill and beyond the sunset, on the province of the American future. There has never been another so excellent or so influential." That report not only charted a road to the Pacific but also added the authority of a land traverse across Oregon country to the claims of ownership asserted by Captain Robert Gray when he discovered the mouth of the Columbia in 1792. As already noted, it ensured the addition of a quarter million square miles to the expanding boundaries of the United States.

But it is the journey itself that continues to capture the imagination. After Lewis and Clark, the uncharted rivers, unprobed mountains, unmeasured plains, and unencountered peoples would never again be so remote and exotic, but the drama of the adventure still inspires an annual assortment of history buffs and neo-explorers to load up the sport-ute, the mini-van, the canoe, or the raft and hit the dusty trail—or the muddy stream—or such parts of them as remain essentially unchanged.

There are still many sites of historical importance worth visiting, and there are still sections of open country between the great bend of the Missouri River and the Pacific Ocean that show no evidence of bull-dozer or plow (though wherever and whenever we find such umbrageous lands, we should immediately try to slap on the cuffs and take them into custody). Exactly how much unprotected wild land is left in the eight states traversed by Lewis and Clark is difficult to ascertain and depends on the definition—whether one is talking about pristine roadless areas that should be set aside as designated wilderness or merely compromised portions of public domain that need a better management plan. There is ample opportunity for action at both ends of the spectrum.

But finding a stretch of turf that shows virtually no evidence of human occupation at all, either before the Corps of Discovery or after, is a rare occurrence indeed. The piece that follows is an account of one such place, a 149-mile segment of the still free-flowing upper Missouri, and of a trip I made down it during the summer of 1978 in an attempt to viscerally experience a tiny part of that greatest of American odysseys.

Bob Singer is a retired high school band director. Now he runs river trips out of a sleepy little north-central Montana town called Fort Benton. There isn't much there anymore since the reclamation dams that begin some 300 miles east toward the North Dakota border delivered the coup de grace to a dwindling steamboat traffic, and the railroad that was proposed to run east–west along the Missouri got relocated to the north along the Milk. This is big sky country with a vengeance. Unbroken horizon. High plains grazing land. Ranches with people scattered few and far between. Fort Benton is the town they visit when they don't want to drive all the way to Great Falls, 40 miles away. It is also the outfitting point for the only remaining free-flowing section of America's longest and historically greatest river, the gateway to 160 miles of a wilderness virtually inaccessible except by foot or boat, the portal to a week of solitude in the last untouched, unaltered, unimproved stretch of the wide Missouri. And that is why we have come.

Singer not only runs river trips, he rents canoes and equipment and hauls people to and from the few ingress/egress points downstream. He is lean, wiry, weather-creased, a human smokestack, a one-man historical society, and if he has an opinion about the five disheveled fools and their 2,000 pounds of junk that he drops at Coal Banks Landing, a few miles below the Virgelle cable ferry, he doesn't express it. No doubt he's used to middle-aged hysteria at the prospect of great adventure; recycled Lewises and Clarks falling all over themselves to be off and away. He helps us unload the canoes from the van, admonishes two little girls trying to stone a rattlesnake that has taken refuge under their fisherman daddy's pickup, promises to meet us six days hence at the Kipp Park Bridge where Highway 191 crosses the Missouri on its way to Malta, and then leaves us to figure out how to get sleeping bags, tents, cooking gear, a week's grub (most of it in cans), and a couple of gallons of water into two 17-foot Grumman canoes.

My old friend and bowman on this expedition, Peter Nabokov, recalls that the ground on which we stand—(he sits, we stand)—was the site of Lewis and Clark's campsite on June 2, 1805, and as the heaviest of our supplies are being lugged down to the river he reads to us aloud from the *Journals*, which we have brought along. "Killed 6 Elk 2 buffale 2 Mule deer and a bear . . . the bear was very near catching Drewyer; it also pursued Charbono who fired his gun in the air as he ran . . . Drewyer finally killed it by a shot in the head; the (only) shot indeed that will conquer the farocity of those tremendious anamals."

"In the event you gentlemen have forgotten," says Nabokov, "Drouillard [né Drewyer] was killed up at the Three Forks by some Blackfeet about four years after this incident. They scalped him."

By mid-afternoon we are loaded and still showing an inch of freeboard. We launch in windless, 90-degree heat and float side by side for a time while we share some cheese and rye crisp and apples. Low bluffs drop steeply to the river along here, and above them the rolling prairie fades off in dun sage contrast to the green ribbon of cottonwoods and willows that screen the bank. In the shallow draws between the hills, little bluestem and bunchgrass and western wheatgrass bow in a breath of hot wind and are still again. Far to the south we can see the sharp etching of the Little Belt Mountains against the bowl of a cloudless sky. We are not far, my bowman informs me, his paddle comfortably tucked in along the gunwale, from the spot where Captain Lewis ascended from the river on June 13, 1805, and looked over what he described as "a most beatifull and level plain of great extent or at least 50 or sixty miles; in this there were infinitely more buffaloe than I had ever before witness at a view."

If we should follow Captain Lewis's example and climb out onto the flats, we ought to be able to see another outlier range to the northeast—the Bearpaws, where Chief Joseph's Nez Perce were cornered on their fighting retreat toward Canada in 1877. As a boy on a Saskatchewan homestead during World War I, my father used to stare southward at the Bearpaws across 50 miles of withered grass, gopher mounds, and heat waves, and dream of cool streams. In the heat it does not seem worthwhile to climb out and return his look. In that direction the map is empty for at least 100 miles, except for a string of little towns along the Milk River. Havre, the biggest of them, regularly competes with International Falls, Minnesota, as the coldest spot in the United States. Today, we suspect, it might be one of the hottest. And we would see none of the buffalo that Captain Lewis saw. Neither would we see any of the Honyaker homesteads that Jim Hill promoted for the sake of his railroad at the turn of the twentieth century. The shacks have weathered and blown away, the fences are occasional leaning posts and scraps of rusted wire, the fields have gone to weeds and inferior range, and the antelope have reclaimed them. The antelope and the Indians. Up in that emptiness are the Rocky Boy and Fort Belknap Reservations, poor farms contemptuously given back to the original owners when white men found them uninhabitable. Until oil or gas or coal or uranium is discovered there, Indian tenure is safe.

Reluctant explorers, we are made vicarious natives by

history. We are not far from the place where, in 1864 or 1865, the Crow warriors Two Leggings and Sews His Guts surprised some Piegan warriors and paused on their journey to the trading post "on the upper reaches of Big River" (Fort Benton) to exchange hostilities. Two Leggings's account of his first coup (in a narrative edited by Nabokov and published by Thomas Crowell) tells of the Piegans firing their muzzle-loading rifles at the Crows and falling back to reload. "One hung behind and I shot him in the shoulder," says Two Leggings. "Reaching back, he jerked out the arrow, broke it, and threw it on the ground. He pulled out his knife and ran at me." Two Leggings shot him in the chest. That arrow, too, the uncooperative Piegan pulled out, broke, and threw on the ground. "I tried to keep out of his reach, yelling to get him excited. Then I shot a third arrow into his stomach. He made a growling sound, but after he broke that arrow he made signs for me to go back. I made signs that I was going to kill him. Then he made signs for me to come closer so he could fight with his knife, but I made signs that I would not." Since his enemy was nearly dead and there was no longer any reason to be cautious, Two Leggings dallied with him, shooting him a time or two for good measure, until finally the Piegan got the message and died while walking back toward his friends. "Then I scalped him and tied the hair to my bow."

Things are quieter along the Missouri now. I let the canoe drift with the current, trailing my paddle in the mocha-colored water. It is so rich with the silt of the unstable soils through which it flows and so alkaline that it is undrinkable, though after an hour in this late afternoon sun a hat full of it feels good on my smoldering pate. Our map, a revised and updated version of one originally published in 1893 by the Missouri River Commission, shows little change in the cut of the channel, little change, indeed, since the upper Missouri was diverted southward during the last glacial advance, and I am reminded once again that the topography before us is almost exactly the same now as it was when Lewis and Clark came through here 175 years ago. When Karl Bodmer, the Swiss artist with Prince Maximilian's 1833 expedition, painted these walls and bluffs and spires, he painted them precisely as we see them now, though he had a habit of foreshortening his scenes.

It's difficult to imagine those first explorers on this river. It's one thing to float down with a strong, albeit gentle, 4-mile-per-hour current behind, but what must it have been like towing a 55-foot keelboat that drew 3 feet of water—against this current—towing it in moccasins and up to your waist in mud? In reference to the stretch of river we are about to enter, Lewis talks about obstructions so continuous that his men are much of the time up to their armpits in the water, and the mud "so tenacious that they are unable to wear their mockersons, and in that situation draging the heavy burthen of a canoe and walking acasionally for several hundred yards over the sharp fragments of rocks which tumble from the clifts and garnish the borders of the river; in short their labour is incredibly painfull and great, yet those faithful fellows bear it without a murmur."

We make our first camp a mile or two inside a section of the river called White Cliffs. The walls here, cut by centuries of flood and fabulously eroded by wind and storm, are nearly perpendicular and vary in height from 200 to 300 feet. Composed of soft sandstone, they have weathered into an architectural symphony of columns, spires, pedestals, flying buttresses, and alcoves. Prince Maximilian in 1833 saw "pulpits, organs with their pipes, old ruins, fortresses, castles, churches with pointed towers." Meriwether Lewis in 1805 imagined "elegant ranges of lofty freestone buildings, having their parapets well stocked with statuary; collumns of various sculpture both grooved and plain." Across from the low promontory above the river where we pitch the tent, the canyon rises nearly 300 feet, and from the dark shadow in a split just below its escarpment the rock swallows that have nested there by the thousands flicker up into the twilight sky like bats out of the mouth of a cave. A flycatcher peers at me with evident curiosity from a dead cottonwood behind the tent, cocks its head, and hops around the backside of a limb to peek at me again, upside down. And when I descend to the river on a sweep for firewood, a great Canada goose lifts out of the marsh grass at the edge of a gravel bar, soars across the water toward the sandstone cliffs, and then veers upstream, climbing steadily until it is lost in the aura of the setting sun. Night

comes down with a pair of mourning doves softly call-
ing from the trees below the camp.

We turn in early and I lie watching the night sky
through the open tent flap, great thunderheads creep-
ing up from the east on a full moon, monstrous anvils
of silver-gray cloud, black-hearted and shot with lightning, beginning to pile up into a space so littered and
bright with stars that even when the moon is finally
eclipsed I can make out Nabokov, cocooned in his bag,
snoring peacefully in the aftermath of his strenuous
day. I roll on my back, thinking about the river, wild,
virtually unmarked, missing only the roving bands of
Minnetarees, Blackfeet, Assiniboines, the buffalo, and
the grizzly bear. Which is a lot, actually. The frontiers-
men may have tamed this wilderness, made it safe for
the likes of us, but at the same time they profoundly
diminished it. That should be some kind of lesson.

Early the next day, about a mile or so above Stonewall
Creek, where the river narrows through a steep-walled
channel, we are swept into an eddy by the current and
quite by accident spot a large nest high up on the cliff
face with three or four downy fledglings peering at us
over their barricade of twigs and sticks. We nose the
canoes into the shelf of mud at the base of the wall and
clamber out, cameras at the ready, to find an avenue of
ascent for close-ups. Let's ignore the suggestion that
we float quietly by and not disturb the natives. Nabokov wants a shot from above, and he is halfway up
a chimney that will take him to the top when a big

red-tailed hawk appears 100 yards downriver, screeching like a banshee and swooping toward us as fast as her wings will propel her. In a matter of seconds her mate appears directly overhead, and together they circle the nest, shrieking their outrage and dive-bombing Stemplepeter, who changes his direction and his mind about baby hawks.

We proceed southeast with measured stroke, led, it seems, by two great blue herons who lift from their mud-shoal perch as we approach, flap a time or two, retract their ungainly legs, and settle into a long, graceful glide that carries them a quarter mile downstream, where they wait for us to come loafing along. Mallards appear in the side channels. There are killdeer, tanagers, and a profusion of magpies in the brush along the cutbanks, and just before we stop for lunch we round the point of a low, treeless island and startle a pair of double-crested cormorants into flight. In real life I am not much of a watcher of birds, but their presence and variety in this wilderness is so insistent that even an apathist cannot remain indifferent for long. I find myself paddling hard to catch up with the other canoe because its occupant, Bob Lewis—no relation of Meriwether—is an ornithological encyclopedia, and he also has a bird book in his pack.

We stop again at the ruins of an old, stone-walled cabin, once a sheepherder's camp, no doubt, now a couple of standing walls with the ax-hewn window headers still intact and a pile of scree where winter snows have toppled in the rest. There are rusted bedsprings and the remnants of a cook stove. Weeds and some small white flowers in the chinks between the mortared walls. Our map tells us that close to this spot, Maximilian and Bodmer camped in the summer of 1833, and we walk the canoes down to a point that resembles one of our reproductions of the Swiss artist's paintings and sprawl on the bank, lunching on half a salami and some rye-not-so-crisp. A short nap; then back on the river.

By late afternoon the country has opened up once more and the cliffs, though still steep and high, have fallen back from the main channel as much as half a mile, the sandstone now capped with a darker rock more resistant to weather so that isolated columns begin to appear that look like grandma's cookie plate, ceramic salt shakers, busted toadstools. The herons seem to like them as observation platforms. The wind has died, and it is extremely hot. Nabokov confirms his growing reputation for careful and considered action by suddenly standing up in the bow and levitating himself over the starboard gunwale into the river. He emerges whooping and blowing with only his idiotic head visible above the turbid water, and enjoins my idiotic son to follow, which nearly overturns the canoe, and I curse them sullenly for fifteen minutes until they begin to complain of the cold and whine to come back aboard. More buoyant than they, I spin off with a flick of the paddle, and let them float into camp.

Day three. Tang and oatmeal bars for breakfast and some salty words to cook Lewis about his expansive cuisine. He wants an early departure, he says, because we will stop tonight at the Judith, a long 25 miles downriver. On the twenty-ninth of May, after a hectic night during which a bull buffalo ran full tilt through their camp ("within a few inches of the heads of one range of men as they lay sleeping"), Lewis and Clark came upon a small but significant stream flowing into the Missouri from the south, and Clark called it the "Judieths" after his intended, Julia (Judy) Hancock of Virginia. They found the campfires of 126 Indian lodges belonging to the Atsina, allies of the Black-feet—or so said "the Indian woman with us" (Sacaja-wea) after she examined their "mockerson" tracks in the sand. Ten miles upriver they came upon a pish-kin, a place where the Indians of the upper Missouri, before they had wide access to the horse and the rifle, used to drive great herds of buffalo over a precipice. The most active and fleet young man of the tribe would put on a skin with the head and horns still intact and position himself so that when the hunters appeared he could pop up and run like mad toward a chosen cliff. The startled buffalo would blindly fol-low; the decoy would try to time his arrival at the edge precisely, drop into a pre-scouted crevasse just below the lip, and let the herd thunder over to its death. Voilà. The most active and fleet young man became the toast of the tepee, the life of the evening party—unless, as sometimes happened, he turned out not to be fleet enough and the thundering herd made a rug out of him.

Lewis remarks that the rotting carcasses they discovered "created a most horrid stench," and that the great many wolves they saw skulking around the neighborhood "were fat and extremely gentle." Clark killed one of them with a short pike called an espontoon, though Lewis does not say why. Probably for the same reason mountain men killed everything—sometimes because they wanted to eat it, often just because it moved. The expedition leaders, in consolation for their unappetizing find at the pishkin, apparently thought it proper to open the bar. Lewis reports that "notwithstanding the allowance of sperits we issued did not exceed ½ (jill) pr. man, several of [the men] were considerably effected by it; such is the effects of abstaining for some time from the uce of sperituous liquors; they were all very merry."

The upper Missouri below Great Falls is strictly a class I river. We look hopefully at markings on our map like Pablo's Rapid and Dead Man Rapid, but they prove to be little more than riffles and a slight acceleration in the current. Just below the Slaughter River at the end of the White Cliffs, we drift down on a flock of white pelicans bobbing in the water, much larger than the coastal variety I am used to, and possessing the greatest wingspan of any bird in this part of the country. We hold our paddles quiet and they watch us until we are within 20 or 30 yards, then lift laboriously off the surface and begin slowly to circle, gaining altitude an inch at a time like some jumbo jet corkscrewing its way out of a high-altitude airport ringed by mountains. They are snow-white with a fringe of black aileron on the trailing edge of their enormous wings, and the conical helix of their upward spiral against the flat blue sky is completely hypnotic. Around and around they wheel, now catching a thermal updraft and rising faster, higher, higher, higher, a speck of cosmic dust in a glinting bank around the sun, still higher—until magically they vanish, absorbed into the atmosphere.

I hear children swimming. The sound of their splashing carries a mile up the river, destroying the midday quiet, the contemplative float through the post-luncheon nods, the illusion that we are alone in this wilderness and protected from the babble of other anthropoids. Somebody is swimming in our river. But nothing appears. The noise grows louder. We float down, almost abreast. Still nothing. I stand cautiously and peer over a low bar of cobble that forms the bank to my right and beyond it into a brackish pond created by summer's recession in the flow of the river. What I see are the glistening backs of a dozen giant carp, their dorsal fins and the scaly upper half of their goldfish bodies cutting through the shallow water in a slow, almost meditative glide, punctuated every now and then by a violent convulsion that churns the water and produces the sound we heard away up river. We beach the canoes and watch them for over an hour, understanding finally the obvious pattern to their ritual. Behind each female (and they appear to be about

3 feet long) swims a smaller male, and it is he who thrashes the water with his tail to excite the releasing of the eggs. The whole scene is so primeval that we would stand gaping all afternoon if the westering arc of the sun didn't jog us on our way.

We camp at the Judith as planned. There is an old ferry here, and a broad grassy bank along the river shaded by cottonwoods. A dirt road runs north for 44 miles to the town of Big Sandy on Route 89, and south from the other side of the river about 49 miles to Hilger on Route 191 out of Lewistown, but neither is a road that anybody travels for pleasure, indeed for any reason except to get back to the ranch. A sign by the ferry says JUDITH LANDING RECREATION AREA, but I don't see anybody recreating except two fishermen setting trotlines for catfish and periodically emerging from their pickup to check on their luck.

It is hard to imagine this empty stretch of river ever changing, ever becoming a "recreation area," with all of the hideous implications of that designation. And not just because it was finally protected in 1976 when Congress, prodded by Senator Lee Metcalf of Montana, placed 149 miles of it in the National Wild and Scenic Rivers System. It is exceedingly remote to begin with. It is not the kind of country that stimulates most backpackers; it has no whitewater to lure river runners; it has limited access and primitive facilities at only a few places. Fishing is pretty much limited to goldeye, sauger, and catfish, and in a state full of blue ribbon trout streams, who cares for that? If

the BLM, charged with developing a river management plan, will leave it alone, it might go unnoticed
and survive. Or if the plan includes banning the use
of motors on boats, dynamiting huge craters in the
primitive access roads, and allowing only ten people
a month (for three months) to float down in nonalu-
minum canoes, gagged and handcuffed to the thwarts,
then it will certainly survive.

But it is a little worrisome when a government
agency is asked to develop a management plan for
anything, particularly when one considers what happened to the rest of the Missouri, the 93 percent under
the direction of the Army Corps of Engineers and
the Bureau of Reclamation. In the 1940s they came
up with two plans for development—the Pick Plan,
backed by the corps, that stressed flood control and
navigation, and the Sloan Plan, backed by the bureau,
that stressed irrigation and hydroelectric power. The
two agencies fought so effectively for their pet projects
that a frustrated Franklin Delano Roosevelt decided
to create a third agency, the Missouri Valley Authority, that would override the first two and get something accomplished. Then the bureau and the corps
decided to cooperate; indeed, they decided to build
all the dams proposed by everybody. And that was
the end of the Missouri River. The economic benefits
from the watershed dams went mainly east, and the
social costs were borne by the people of the regions
through which the river flows—like the Mandans,
who lost much of their cultural history beneath the

waters backed up by the Garrison Dam. So much for management.

The fishermen are at their lines in the morning when I walk down to the river to splash water on my face. We exchange pleasantries. One goes back up to look after breakfast; I stand with the other admiring the view, letting the morning sun warm my bones and ease the stiffness from a night on the ground. "You from these parts?" I ask.

"Born and raised," he says, baiting a hook. He has a furze of Gabby Hayes whiskers and a rooster tail of gray hair poking up where he slept on it. "My daddy came out to this country in 1890 and homesteaded a piece down on Hell Creek. Underwater now. The Fort Peck dam flooded it."

"That's a real shame."

"Not much it ain't. I'd probably still be stuck there if it weren't for the dam." He ties his line to a short stick and jams it into the mud, throws the baited hook into the river, and we watch the current sweep it down. "Live over in Idaho now. The wife and I are just back to see her people in Lewistown, so Russell and me"— he thumbs backward in the direction of his departed companion—"came fishing."

"So, do you miss it any? This country, I mean."

"Oh, it's pretty nice country. But it's hard. I can remember before the war I used to work summers to get me a stake, buy myself a new rifle and some traps, and take off south along the river here to winter over.

I could make a hundred dollars trapping for the season. In those days it was better than working for keep. Where you all from?"

"California." I am sincerely hoping we won't have to go into it. "What did your father do down there on Hell Creek?"

"Well, he raised horses mostly. He'd catch mustangs lost by the Indians or the cavalry and use them for breeding stock." I look attentive through the pause that follows, and he opens up a little. "He had a line that was half Tennessee walker and half mustang that we called a Tennessee Whip, because the way you broke it was with a whip. Only way to tame it was to beat the hell out of it until you got its attention, and then I'll say it was a purty good horse . . . people prized it above any in these parts. But it was so Goddarn mean it'd kill you if you didn't stay awake."

He brushes the sand off the seat of his pants and lowers himself onto a patch of grass, taking a can of Copenhagen out of his shirt and offering me a chew. "Whereabouts did you say in California?"

"Near San Francisco. Did you ever have one of those horses yourself?"

"No sir, I didn't, but I was just a pipsqueak then. My daddy had one he'd whip broke that he couldn't ever let it see both of his hands at the same time. If it only saw one it figured daddy had his whip in the other and if it messed up it was gonna get coldcocked. But if it saw both hands, well, then it would try to stomp him for sure." He spits a few crumbs of snuff between his rubber boots and reaches down to pull in his line and remove a branch that has snagged in it. "I remember one old boy he hooked up with down there at Hell Creek, a trapper I think he was, and he had him one of those horses that had both its ears chewed off. That was how he'd learned it to pay respect."

From back in the cottonwoods I hear his companion holler to come and get it, but he is evidently a little deaf and makes no move to get up. "It was quite different, this country, back then," he says, folding his hands around one knee and rocking back. He appears to think about this for a while.

"Fewer people?"

"Oh my, no. Used to be all the land up the river to Fort Benton was homesteaded by folks who thought the railroad was going to come through. When they put it up along the Milk, most just pulled out. Moved into towns. Now that's another thing. Towns. Towns ruined this country. You know, it used to be that three or four sections would get together, three or four homesteads, and they'd hire a schoolteacher from Oklahoma or somewhere and set up a community school for the kids. Then the government started trying to consolidate everything into the towns, I don't know what for, and they began paying eight dollars a kid if their folks would transport them in to the town school." I hear his friend bellow, "Soup is on, damn it."

"That's when everything went haywire. Because, you see, if a rancher had four kids, why that was thirty-two dollars, which was enough for his wife and family to move to town and rent a place for the winter. Of course, the old man stayed out there on the place alone, and after a bit that got real lonesome, and the next thing you know he'd moved into town too, found something he could take up to do, and as often as not he never went back. It was easier in town. In my opinion, that's what ruined this country."

From the bank above, "Goddamn it, William, you deaf old sucker, the flapjacks are gettin' cold. Fact, they are cold."

"Well, it's been nice talking to you," I say. "Good luck with the fishing."

William rises arthritically to his knees, then to his feet. "You enjoy the rest of your trip," he says. "It's purty nice country."

Below the Judith it is less nice. Or at least I think it not so interesting. Others who have written about this river have found the badlands along this eastern stretch more teeming with wildlife than the White Cliffs section, and the closer one gets to the Charles M. Russell National Wildlife Range, the more this ought to be so. Bighorn sheep and elk have been reintroduced in the area. There should be whitetail and mule deer in abundance, and this is the most likely area to spot a golden eagle. But we see, in fact, fewer birds, more signs of human habitation, and apart from a small herd of pronghorn and a few beaver, not much along the banks. Captain William Clark, who more and more I am inclined to think was a distant relative of Machine-Gun Kelly, suffered no such disappointment. He remarks in his journal entry for the twenty-third of May, "I walked on shore and killed 4 deer & an Elk, & a beaver in the evening we killed a large fat Bear, which we unfortunately lost in the river . . . The after part of this day was worm & the Musquetors troublesome. Saw but five Buffalow a number of Elk & Deer & 5 bear & 2 antilopes to day." The following morning, just to get his blood circulating, he "walked on shore and killed a female Ibi or big horn animal . . . in my absence Drewyer & Bratten killed two others." A kill count from the Lewis and Clark journals might help explain our diminished sightings.

The weather begins to turn foul, with high winds and blustering rainsqualls, and because we laid over a day at the Judith to swim and loaf, we begin to feel pressed to make time. We aren't, but that doesn't seem to occur to anybody. Nabokov suggests we lash the canoes together and rig a sail out of a ground cloth and two paddles, and he and my son sit in the bow holding this contraption before the wind. No wonder we don't see any wildlife. We look like the last days of the *Kon-Tiki*, veering and yawing down the river, flapping like wash day at the asylum, but it works, and we rocket through 35 miles of twisting channel before we finally collapse at a place called Bullwhacker Coulee. Meriwether Lewis climbed out of the river here on the twenty-sixth day of May, a little over a year and 2,000 miles from the day he and Clark and forty-three men began their trek to the Pacific, and he caught his first glimpse of what he thought were the Rocky Mountains, "covered with snow and the sun shone on it in such a manner as to give me the most plain and satisfactory view." His reaction to this discovery is all the more touching because he was actually looking at the Little Rockies of northern Montana, rather than the great mountains he thought them to be. "I feel a secret pleasure in finding myself so near the head of the heretofore conceived boundless Missouri; but when I reflected on the difficulties which this snowy barrier would most probably throw in my way to the Pacific, and the sufferings and hardships of myself and party in them, it in some measure counterbalanced the joy I had felt in the first moments in which I gazed on them; but as I have always held it a crime to anticipate evils I will believe it a good comfortable road until I am compelled to believe differently."

I wonder what he would think of that road now? Take Interstate 15 to Helena and over the Continental Divide to Butte. Pick up Interstate 90 to Missoula and then Highway 12 across Idaho and into Washington. You'll hit the Columbia River just past Walla Walla, and from there it's interstate again all the way to the ocean. Six months, you say? Now it's a two-day drive.

Rain continues to fall in windblown sheets, and we turn in at the first hint of dusk. In the morning we load the canoes and shove off in a gusting storm that turns the river to chop. Cow Island, where Chief Joseph and his band of Nez Perces crossed in September 1877 (General Oliver Howard in hot pursuit), slips by to our left. In late summer and autumn when the river is low, it was the headwaters of navigation for the steam-boat traffic pointed west, and it was here that the supplies for the Royal Northwest Mounted Police were stored until they could be hauled by bull train north to the Canadian posts. The channel is much wider along here than it was above the Judith; the cliffs are far back from the muddy banks, low and unremarkable. There are numerous side channels and grassy, tree-less islands. Again, perhaps because of the weather, we see little in the way of animal life except cows and an occasional band of horses, and few birds except a crippled pelican, obviously the recipient of some idi-ot's shotgun blast, dragging itself along the beach at Tea Island, just before our pullout point at the Fred Robinson Bridge.

Civilization. After the wildness of the upper river, the solitude, the illusion of being sequestered from humanity, returning to a world even as sparsely inhab-ited as northeastern Montana suddenly begins to seem like a bum idea. We begin to regret out loud that 35-mile day when we did little but hunker behind an improvised sail. We should have spent more time off the river, hiking, poking around. We start to ques-tion one another about the infernal compulsion to log miles, check maps, and figure out where we are and how far we have to go. Go where? To a steel and con-crete bridge with cars zipping overhead, overflowing refuse cans, chemical johns, wounded birds? Pretty soon we aren't even speaking.

But it is only by contrast that this feels like civilization, and it seems extremely unlikely that this piece of the world is going to be much changed by an influx of tourists, developers, in-migrants, speculators, retirement communities, second-home owners, utility companies, or resort operators. There just isn't anything up here to make a buck on. Space and the wind in your ears. No thrills. For most people it is too hot in the summer, too cold in the winter, too austere in its emptiness, too far from port, too tough on the spirit. It will remain, one feels, the preserve of those who can accept it on its own terms.

Nuestra Señora
de los Dolores

*Snaggle-tooth (snag'el-tooth') n. 1. A tooth that is broken
or not in alignment with the others. 2. A rapid on the
Dolores River. 3. A particularly vicious, mean-spirited,
ugly impediment in the lower-right-hand corner of what
appears to be a dramatization of organic brain disorder.*

THE HANDBILLS will tell you there's something for everybody on El Rio de Nuestra
Señora de los Dolores—the river of our lady
of sorrows—although they skip the translation and
throw you sucker punches instead, like photographs
of spring wildflowers, white yarrow, aster, penstemon,
and purple daisies, snowcapped mountain peaks, cliffs
of Wingate sandstone, Anasazi rock art (commingling
with Great Basin redneck rock art), zonal transforma-
tions. They'll suggest that if Douglas fir and yellow
pine are not your cup of tea, you have only to wait for
the lower elevation piñon-juniper forests; and if you
don't like piñon-juniper, wait for the sagebrush and
rice grass.

There are indeed photo-ops around every twist of

the Dolores canyon, but let us not be deceived by slick brochures put out by adventure companies that always show their customers (known in the trade as "peeps") grinning broadly at the camera, helmets askew, paddles at the ready, utterly ignorant of the 10-foot plunge they are about to take over a blind fall just beyond the camera's range, down into a vicious boulder maze that leads straight into the most horrifying, boat-flipping hole ever encountered—or at any rate the most petrifying, boat-flipping hole ever encountered by the "adventure consultant" (actually a trainee) who has mistakenly been designated the trip leader for this voyage. These handbills show only the "before," never the "after"—never the carnage scattered up, down, and sideways across the river; the peeps clinging for dear life to midstream boulders; the adventure consultant tangled elbow to ankle bone in the flip lines of his or her overturned raft and about to be swept into the lethal arms of a downed cottonwood strainer.

They fail to mention the steep gradient of Nuestra Señora, the fast spring runoff, the absence of eddies into which to pull a boat when one needs to put ashore, the rattlesnakes, the icy water, the rain/hail/snowstorms at higher elevations, the bugs and ferocious heat below the Paradox Valley. They don't refer to river contamination from the numerous uranium mines in the region, so toxic that the redoubtable eco-terrorist James Watt, secretary of the interior during the reign of King Ronald the Popular, thought it would make a terrific site for a spent fuel rod dump.

And they don't tell you the reason why that miserable viper of a rapid they refer to in ominous terms as the "infamous snaggletooth" is, in fact, infamous. For good reason is this river named, in memoriam, our lady of the sorrows.

Nevertheless, regardless, and notwithstanding, we rendezvous in Monticello, Utah, and drive east along bad omen Highway 666 to the misspelled town of Cahone, a two-bit hamlet in the southwestern corner of the vastly overrated state of Colorado, thence down 6 miles of treacherous and often vertical dirt road to a put-in camp by the river. The Dolores actually begins some 60-odd miles upstream in the San Juan Mountains near Lizard Head Pass and flows more or less west/southwest before being impounded by the McPhee Dam at Cahone. The upper reaches above the dam can be run, but only by kayak fanatics obsessed by narrow, *pínche* little streams full of nasty rocks and unholy drops. Below the dam it widens as it doodles and dangles north for 178 miles, dropping some 2,500 feet from its origin to its confluence with the Colorado River near Arches National Park. Out of perversity, one supposes, it runs northwest across Disappointment, Big Gypsum, and Paradox Valleys, despite the fact that the stratigraphic gradient of the Uncompahgre Plateau, which it flanks, runs southwest. There is no simple explanation for this anomaly other than that the topography was clearly different 50 million years ago when La Señora began her sorrowful trek in search of Lake Uinta, that prehistoric pond formed during the late Tertiary Period to fill in the basin between the Uinta Mountains on the north, the Wasatch on the west, and the Roan Plateau and Book Cliffs to the south.

But we plan a less ambitious voyage, just over 100 miles to the highway crossing at Bedrock in the Paradox Valley, so named because of the glitch described above—the river runs across it instead of along its long southwestern axis. At the Bradfield launch site there are fifteen of us in four oar-powered rafts: a fresh batch of California college students on a "field study"; four "adventure consultants" to minister to their culinary needs, domestic arrangements, and libido crises; and one tagalong, whom we find sitting morosely on his river bag at the put-in. The commercial trip on which he had been booked had shoved off early, leaving him in the port-a-pot, where he had sequestered himself with a joint and a copy of *Outside* magazine.

During the safety talk, wherein Adventure Consultant Bud Bogle imparts to the peeps the imperative use of life jackets at all times, the feet-first, on-your-back position for negotiating a rapid should one find oneself prematurely disembarked, and the proper handling of throw ropes, Tag confesses that he can't swim. "The deepest water I've ever been in is my bathtub," he confides, a record that will soon be broken in the run-out below the Snaggletooth.

In its 200-mile journey to the Colorado River, the Dolores drops through four major Western life zones, from alpine at its headwaters to the Upper Sonoran along its lower reaches. At the Bradfield put-in, and during most of a very short first day (due to a late afternoon start we float only a few miles), we are in the transitional zone, basically open ponderosa pine forest, though there is some evidence of Gambel oak and a few aspen common to the higher Canadian zone. Thirty miles downriver, assuming we survive, we will have dropped into the Upper Sonoran desert, with its scrub environment of piñon-juniper, chaparral, buckthorn, manzanita, and sage.

The ponderosa flow by against their backdrop of high cliff, and the current, though swift (an average drop of 19 feet per mile), is without serious challenge. No surprises here. The river cuts north/northwest through a 2,000-foot-deep canyon it has carved into the Dolores anticline and as we bob along between its banks, on four overloaded baloney boats of well-patched hypalon, we keep our eyes peeled for bighorn sheep and mule deer. Cougar are said to inhabit this terrain, but we don't see any of them either. In fact, our only wildlife encounter is a pair of collared lizards doing pushups on the warm stone of a huge boulder we slide precariously close to around a sharp bend in the river.

The students abandon ship the moment we touch shore and disappear to set up their nests, ignoring in the process an earlier and obviously too mild explication of river-rafting etiquette: *first* we unload boats, *then* we set up kitchen and port-a-pot, and then *and only then* do we attend to our personal space. The food allergies and nutritional obsessions that have become de rigueur among California undergraduates surface immediately, and at the dinner trough there are the usual complaints about beans and pork by-products, non-organic produce, aspartame, phenylal-anine, sodium nitrite, transfats, not to mention the complete absence of soy milk, hummus, edamame, Weet-Bix, barley grass, blue-green algae, and so on. Certainly no shortage of hydrogenated hubris among these maggots, Bud observes. But somehow we get everybody fed.

Our second day is as uneventful as the first, except for a disagreeable incident at breakfast when a dispute erupts between two girls over the ownership of a green plastic plate. Because each member of the expedition has been required to provide his or her personal kit, the loss of one of its component parts, particularly its centerpiece, is a matter of consequence. Both women lay an increasingly noisy claim to this item of syn-thetic crockery, tugging and pulling it back and forth between them, growing moist, agitated, and increas-ingly profane—until eventually they upset the coffee pot from its perch on a log and bring adult interven-tion down upon themselves.

"So how are we going to settle this?" Bud inquires, stepping in and taking custody of the plate. "Shall we go halfzies?"

More sputtering, clenching, and unclenching, spit-tle now flying.

"Ladies?"

"Fuck off, Bud," one of them snarls.

"Ah, so it's 'Fuck off, Bud,' is it?" Bud says, regarding them for an impassive moment. "Well, okay . . . that's cool." He turns toward the river and Frisbees the plate toward the beach on the other side. The assembled watch it deflect off a boulder, sail high over the bank, and disappear into a dense thicket of tamarisk. The Solomonic wisdom of this resolution leaves both combatants utterly nonplussed. Bud offers a wisp of a smile and gestures at the upturned coffee pot, clearly inviting the perps to clean up the mess.

And then, of course, at the evening meal the vegans are joined by two wheat allergies and an *E. coli* victim and skip the roll-your-owns (flour tortillas probably containing shortening or lard or some artery-clogging substance—never mind the meat and cheese). The chocolate brownies baked in a well-greased Dutch oven are consumed by all parties, even though they contain unspeakable ingredients listed on the pre-mixed box. It is noted that we are still carrying a full complement of diet sodas, rejected, one chucklehead asserts, because they contain phenylalanine. Whatever that is.

After cleanup chores we lounge about in the fading light, listening to one of the students pick the guitar and entertain us with a lengthy version of "Geronimo's Cadillac":

> Hey boys, take me back,
> I wanna ride in Geronimo's Cadillac;

oh boys I wanna see it for real,
I wanna ride in Geronimo's automobile.

We turn in early because we have a long day tomorrow, including a run through the villainous Snaggletooth, before which everybody wants to get right with God.

We are camped just below Mountain Sheep Point, a recreation site 6 miles east of Dove Creek and the devil's highway, 666. Our plan is to get an early start, float down to Snaggletooth, scout it, run it, and then stop for lunch at the pull-out below the rapid. Beyond Snaggletooth it's pretty smooth sailing, so we'll try for a late camp some 15 or 16 miles downstream, where the canyon opens out into rolling hills above the bridge at Slick Rock.

But the devil seems to have been hanging out at the Box Elder camp and has sneaked aboard the sweep raft, manifesting his presence at an unnamed, house-sized rock that juts out from the right bank and forces the river into a sudden and unexpected left-hand turn. A boatman needs to execute an abrupt, stern-first pivot and a hard downstream ferry in order to avoid a catastrophe, but the boatman in our sweep boat is asleep at the wheel, and before he can utter "Oh, shit!" he piles up sideways against the immovable obstacle and wraps. Which is to say, his upstream tube is forced down by the current and is plastered against the undercut boulder with only about 2 feet of the bow of the boat left poking up out of the water. The passengers leap to the high side and scramble like cockroaches onto the rock. Gear begins to tear loose from the submerged frame—duffel bags, the spare oars, extra life jackets, flotation cushions, a dozen oranges, the port-a-pot, a watermelon. There goes a blue, plastic tarp; a torn river bag disgorging T-shirts, shorts, and a bikini; the ill-fated coffeepot; and 10 gallons of propane. Much of this flotsam will be recovered downstream, but it's clear that a lot of people are going to go to bed tonight in wet clothes in a wet sleeping bag.

Roping up, we take turns bouncing on the one visible part of the bow tube, attempting to slowly jar it forward along the rock where it may eventually pop loose. Progress is slow, but as more and more of the tube becomes visible, we begin to take heart. It actually looks promising. Old man river wants to puke this thing up and be rid of it. Bounce, bounce, bounce. More gear floats free. Bounce, bounce, bounce. At last, with a big sucking *fffluuumphh* the bugger comes loose like a breaching whale, and the last bouncer, one Joe Pack, the inattentive slacker who caused this wreck in the first place, does an entrechat, a plié, and topples off the rock into the river. Like a piece of epidermal Saran Wrap, he replaces the liberated raft, plastered by the current against the undercut stone, with only half his upper torso out of the water. This might be grimly amusing, except we can't exactly do a dance on *his* head to jar him free, and because the force of the current is preventing him from moving any part of his body, he is in danger of being forced under and drowning.

We rope up another volunteer, secure his feet in a noose, and lower him face first down toward lonesome Joe. Can you move anything, Joe? No. Try scrunching your fingers. Can't. Try. Slowly, Pack is able to crab walk his arm upward until it is above the water line and able to grasp an extended hand, and inch by painful inch we jerk him upward to safety, unfortunately leaving a considerable amount of him grated like cheese on the rock.

Onward. At this rate we'll be *camping* below Snaggletooth. A moment, however, to set the scene. The entrance to Snaggletooth Rapid is down a tongue leading into a chute on the left side of the river that half fills any boat passing through it. It is 300 or 400 yards upstream of the tooth, a nasty jagged fang near the right bank that must be passed on its right side. All of which means that our fearless guide must immediately undertake to downstream ferry himself, passengers, gear, and his half-swamped boat clear across the river in an exceedingly swift current. If he's lucky he just makes it and can relax his sphincter as he squeaks by. If he dogs it on the oars, however, he either wraps his boat around the tooth or slides off on the *left* side, which dumps him into a severe pour-over where he flips (if he's lucky) or is projected skyward in a spectacular endo (if he's not). Or so says Ann Cassidy, who knows whereof she speaks.

Straightforward, this maneuver, with one caveat. If our boatman is too hyped and Herculean in his cross-stream ferry, he's likely to arrive at his desired entry too early and wind up among the boat-ripping rocks barely submerged along the whole right-hand bank of the rapid. As we are about to see.

Observing Snaggletooth from above, one better understands the dread a trapped fox must surely feel, held in the vise grip of an implacable force that is utterly indifferent to its kismet. Today, at whatever volume the river is flowing, no options present themselves as to the run—we must cross the river, pass the old "snag" on the right, get clear back to the left bank, and land in fast water on a small beach before being swept around a bend into another long stretch of turbulent froth. Which ends in an undercut cliff. Very bad news for swimmers.

The chute through which one enters this incubus is a thundering trough of whitewater that looks like an open spillway during flood peak. The left side of the river is a rock maze of broken boulders and refractory holes all the way down past the tooth, impossible to negotiate, and the slot to the right of that monolith appears to be just about wide enough for half a kayak. I sincerely hope it is an optical illusion caused by angle, distance, and drop.

My customary load of student passengers decide

they'll hike around Snaggletooth and meet me at the bottom. Me? Isn't anyone going with me? I need someone to bail when I fill up in the entrance chute. I need moral support, quislings; I need trust, confidence, love . . . at the very least valium. "I'll go with you, sweetie," says my faithful wife. And old Tag Along, bless his heart and a southern-fried upbringing, will not leave a buddy in his hour of need. "I'll go with you, man," he says. "What the heck, I need to learn to swim."

Now the only major obstacle to departure seems to be stark terror. My mouth contains the moisture content of a desiccated road apple. The limbs seem to work, though not necessarily in conjunction with commands from above. Lynn and Tag, like kindly guards assisting a beshatted prisoner from his holding cell to the death house ("Sorry, Stegner, it's time.") lead me down to the raft and fasten the electrodes. Silently, they switch on the current.

Which carries us down the long tongue of the rapid and into the exploding flume at its apex. V-waves detonate over the bow and swamp the boat as abruptly as God's own dipper filling a water glass. I retrieve the arm that has been torn out of its socket along with the left oar and begin frantically rowing across the river screaming "BAIL," torquing my neck like a pretzel to try to catch a glimpse of the tooth I know must be charging upriver to meet us. But it seems to have been vaporized, can't see it . . . total whiteout here. "Let's BAIL, people! No time to be dogging it." Ah, the hat's over my eyes, no wonder . . . now that's better, there it is, miles to go, miles, way down river, no sweat, but Jesus I don't want to overshoot this hussy, better ease up or I'll be doing a Mike Walker here . . .

I don't know exactly when it occurs to me that I'm going to hit the Snaggletooth dead on, but the realization has more humor in it than fear. What's that old joke about the last thing to go through a bug's head before it hits the windshield is its mind? Well, things are out of my hands now, kismet in charge, and it's Ann's fault anyway, she got me so nervous about rowing too hard. Maybe she'll be down there with some throw ropes.

My last act as we pile into what I see, up close and personal, is just a gigantic fang of decayed granite, is to pull hard on the right oar and spin us into the facinerous, crushing, outrageous hole on the left side of the rock, where an astronomical back wave performs the Heimlich maneuver on our little hypalon tub, and we are ejected like spattering gobs of steak fat into the great maw of Nuestra Señora.

Flushed out below the rapid after the spin and rinse cycle is completed, Lynn and I swim frantically for the little beach before we are swept around the bend. Tag, however, has thrown in the towel and floats listlessly in his lifejacket, resigned and submissive. He understands now that Mother Nature does not conduct business in a court of law, that double jeopardy is, in fact, her stock in trade. Clearly it is his destiny to drown in the river of sorrows. He hopes he will be fondly remembered.

When the throw bag hits him upside the head, he has no cognitive awareness of its function or purpose. But he grabs it, clutches it to him, and, as he tells us later, finds a certain courage in the developing fury he feels toward the idiots on shore who keep trying to jerk it away. He's damned if he'll give it up.

Thus in anger do we redefine karma.

Let us turn to the outstanding wildlife values along this lush riparian corridor, the aquatic diversity, the multiplicity of avian species, the variety of terrestrial genus. Look at the squawberry and willow, the mule deer and red-spotted toad, the flannel-mouthed sucker. Consider the fossilized dinosaur bones exposed by the unearthing of the Morrison Formation. Remember Father Escalante, who named this damned river of sorrows. In short, focus on anything but the wreckage strewn along the banks below the rapid, and the bedraggled survivors picking their way through what might as well be the great garbage dump of Dingypurna. They will survive. Leave them in their anguish and heartache. Regard instead the side-blotched lizard, the speckled dace, the mottled sculpin.

The rest is the tedium of uneventful floating, the boredom that chaperones tranquility. At Paradox Valley, named for the geological formation, not the contradictions of whitewater rafting, Highway 90 crosses the river, and we will exercise the option of taking out at the community known as Bedrock. The alternative is four more joyless days to the confluence of the Dolores and the Colorado just above Moab, Utah, and nobody is clamoring for that. So there's nothing to do now but settle back and stare at two-hundred-thousand-million-year-old vermilion cliffs of Wingate sandstone, streaked by desert varnish, laced at every seep and spring with hanging gardens of fern and monkey flower, overseen from above by golden eagles and from below by great blue herons as Triassic as the canyons through which they cruise. Even the sun comes out to spoil the ambience.

THE BRIGHT EDGE
OF THE WORLD

D URING THE SUMMER of 1993, I was asked to write a piece on the Colorado Plateau for a special spring edition of *Sierra* magazine that would attempt to define the unique character of some six major eco-regions of the United States, and as I had a simultaneous invitation to join an early fall river trip through the Grand Canyon, I decided to spend some pre-float time reacquainting myself with my subject. My fellow "freelance ecoregionalists," as the Sierra Club described us (Bill McKibben, Andrei Codrescu, Jane Smiley, W. S. Merwin, and John Daniel) all actually lived in the territory they were to expose, and would speak, one had to presume, with authority about their topic. I, however, lived in California, which as we all know is a terrestrial paradise invented in 1510 by Garci Ordoñez Rodriguez de Montalvo, inhabited solely by black, uni-breasted Amazons and Robert Stone's ravagey little raccoons of the mind, and none of its six eco-regions (some say seven) have very much at all in common with the canyon country of southern Utah. I needed to put Tevas on the ground. Do my homework.

Let me concede at the outset that a few thousand words are hopelessly inadequate to the task. The Colorado Plateau encompasses 130,000 square miles of southern Utah, northern Arizona, northwestern New Mexico, and southwestern Colorado and is a physiographic "province" (i.e., a major landform that differs geologically from everything surrounding it) that extends from the Rocky Mountains to the Great Basin, from the Mogollon Rim to the Uintas. Or if one wants to think vertically, it rises from a low point at the 1,200-foot base of the Grand Wash Cliffs above Lake Mead to a high point at the summit of 12,633-foot Mt. Humphreys in the San Francisco Mountains—a zonal ascent from Sonoran to alpine with a broad spectrum of biotic diversification along the way.

In truth, the Colorado Plateau doesn't fit one's conception of a plateau at all. Composed of several miles of stratigraphic layers laid down by the comings and goings of ancient seas, it remained structurally intact while its neighboring provinces to the north and east (Rocky Mountain Province) and to the west (Basin and Range Province) were warping, faulting, tilting, uplifting, leaving it surrounded by highlands like an immense, heart-shaped basin or lagoon. Unfortunately, that image doesn't satisfy either, since it, too, implies something flat, and our basin (née plateau) contains the colossal river canyons of the Green, Colorado, Little Colorado, San Juan, Dolores, Escalante, and Dirty Devil—as well as nearly fifty other major tributary side canyons. In short, ten million years of erosion have cut the province into a myriad of plateaus.

And it goes as high as it goes deep. Stretched out along its southern rim are the lofty volcanic peaks of Mt. Logan, Mt. Trumbull, Mt. Humphreys, and Mt. Taylor. At its center the Henry, Abajo, and La Sal Mountains rise over 11,000 feet in three outbursts of isolated peaks—laccolithic domes, actually, volcanoes that never vented their spleens through the sedimentary overlay.

Mountains and canyons notwithstanding, the region is generically classified as "desert," since most of it receives less than 10 inches of rainfall a year. Desert is about as useless a term as plateau. What kind of desert? Sonoran, Mojave, or Great Basin? (The province contains all three.) And since when are Hudsonian life zones found in a "desert"?

No, it can't be done. Give it up. No geophysical inventory is going to do justice to the territory. We ought to at least give passing notice, however, to a few of the topographical features commonly observed—all those badlands and slickrock mesas, those benches, basins, reefs, rims, dikes, salt valleys, hogbacks, upwarps, monoclines, saddles, buttes, craters, arches, bridges, ridges, cliffs, towers, needles, monuments, rincons, palisades, pinnacles, spires, seeps, washes, creeks, gulches, ravines, chasms, gorges, and dunes scattered about out there, hither and yon. And, of course, we should note that the Colorado Plateau is not just morphology and geology. It is climate, ecosystem, and human history (though only a paltry 12,000 years of it), and for many of its biotic communities, including the hominids, it is threatened habitat. None of these subjects can be reduced to simple definition either. One might as easily synthesize the states of Maine, New Hampshire, Vermont, Massachusetts, Connecticut, Rhode Island, New York, and New Jersey—which in total area is a smaller space than the plateau province.

Robert Frost once remarked that all art is synecdoche, and it seems to me that all one can do in a literary translation of the Colorado Plateau is apply the same principle—punch a couple of test holes here and there, take some samples, let one or two cows represent the herd. Because I am an expatriated son of Utah, born there but, as noted, residing primarily west of the West, it seems prudent to collect my specimens in territory that, as a consequence of a multitude of river trips that have had their inception or termination in the heart of canyon country, I think I know a little about. To that arrogant end I find myself on that fine fall afternoon motoring south toward I-70 on Highway 6 at the base of the Book Cliffs, a 1,000-foot-high, 250-mile-long escarpment (the longest unbroken escarpment in the world) that lies in an east–west arc between Price, Utah, and Grand Junction, Colorado. It is a clear, cloudless day, 80 degrees, and I can see 100 miles to the south where the snow-capped peaks of the Henrys poke through the horizontal seam between slickrock and sky. In the ditches along the highway, the dry stalks of globe mallow blow in cheerless consort with the grey Mancos shale of the surrounding desert.

The Book Cliffs, and the Roan Cliffs behind them, mark the abrupt termination of the Tavaputs Plateau, a gargantuan, 2-mile-thick block of sedimentary uplift that flows gradually southward from the Uinta Basin to the San Rafael Wilderness and the Grand Valley of the Colorado River. It also defines the northern boundary of the "biological heart" of Utah's plateau region—a term I shamelessly steal along with a lot of other stuff put out by the Nature Conservancy's Great Basin Field Office in Salt Lake City. The conservancy has recently launched what it calls the "Bright Edge Campaign" to identify and help preserve specific eco-regions in southern Utah, and in order to concentrate energy and resources into an effective program, it has focused on about 4 million acres that extend east–west along the Book Cliffs from Price to Cisco and north–south from the Book Cliffs to Blanding, where an estimated 75 percent of all plant and animal species on the plateau can be found. What better place to take samples?

In the town of Green River I stop at Ray's Tavern for a coronary thrombosis and a pint of cholesterol. Ray's serves what is indisputably the world's best cheeseburger and fries, half pounders each, fries so waterlogged in grease one is advised to consume them wearing latex gloves. On the wall behind the bar there are action photographs of hairball runs through who-knows-what stretch of rapids on who-knows-what river—certainly not scenes from the local whitewater attraction, the Desolation/Gray Canyon section of the Green River. For over 90 miles the Green has carved its way down into the Tertiary sediments and Cretaceous shale of the Tavaputs Plateau before emerging from a slot in the Book Cliffs to flow mellifluously out past Ray's Tavern and into the San Rafael Desert, but on the oarsman's panic scale it barely registers a 3.

On a wilderness scale it registers a 10. "Crags and tower-shaped peaks are seen everywhere," wrote John Wesley Powell as he passed through it in 1869, "and away above them, long lines of broken cliffs; and above and beyond the cliffs are pine forests, of which we obtain occasional glimpses as we look up through a vista of rocks." The rocks and the vistas have not changed a whit in the 124 years since Powell first saw them, though were he to float the river today he might be surprised to discover guest lodges along Range Creek, a light-duty road through Horse and Little Horse Canyons, oil exploration wells in Tusher Canyon, drill pads on the Beckwith Plateau, a pipeline in Jack Canyon, and an Avon or Domar boat in every good campsite from Sand Wash to Vesey's Rapid.

The flora and fauna have been more affected than the scenery, but the Book Cliffs/Desolation Canyon wilderness is still referred to as the "Serengeti" of the Colorado Plateau because it provides habitat for an incredibly diversified wildlife population—cougar, black bear, elk, deer, bighorn sheep, coyote; bald eagle, peregrine falcon, large numbers of songbirds, ducks, and shorebirds (the Green is a major migratory route); Colorado squawfish, humpback and bonytail chub, catfish, and trout populations in the tributary creeks.

Not to mention venomous reptiles falling from the sky. On my last trip through Deso/Gray with my wife and Bud Bogle, we were bombarded somewhere in the vicinity of Wild Horse Canyon by a young golden eagle that launched itself from a high ledge directly above our raft with 3 feet of writhing, twisting, supremely pissed-off rattlesnake in its talons. And then proceeded to lose altitude. There was a indeterminate period of flapping wings and serpentine squirming before the eagle accepted reality and let go, and the snake came pinwheeling down on top of us. We did not expect much from it in the way of an attitude adjustment when it smacked down on the floor of our hypalon raft, so we bailed . . . as in out. Fortunately the Green, unlike the Colorado, is not a cold river to swim.

A reincarnated Powell would also note that since his day the Tavaputs Plateau has become habitat for a new phylogenetic oddity, the deadly and ruinous *Homo sapiens*, a flat-footed, two-legged, off-road mammal who annually logs 60,000-user days on the river and who comes each season by the thousands to blast away at the deer population. There are a couple of them nursing brewskis right here in Ray's Tavern, sporting T-shirts from the Roadkill Café that announce its daily specials down the back: Poodles and Noodles, Flat Cat (served as a single or in a stack),

Thumper à la bumper. Across the chest it says, "You kill it, We grill it."

In 1936 Bob Marshall, founder of the Wilderness Society, identified the area surrounding Desolation Canyon as the fifth-largest desert wilderness (i.e., roadless area) in the United States, and approximated its size at 2.4 million acres. Over the ensuing years more than 60 percent of the region has been deflowered by coal, oil, and natural gas developers, but it is still one of the largest roadless areas in the intermountain West. Unfortunately, of the million acres that remain, only about 150,000 are protected. And given the relentless opposition to "locking up the wilderness" by Utah's BLM, the prospects for any substantial augmentation are still slim to uncertain.

Two-thirds of Utah is federal property, of which about 22 million acres are administered by the BLM. It is certainly news to almost nobody that the BLM and its counterpart, the Forest Service, have long been dismal custodians of the public lands, but on the off-chance that there is somebody out there who has missed it . . . Disregarding the obvious intention of the pre–Bush administration Congress to save a reasonable amount of what little public domain remains unleased, undeveloped, and unexploited, both the Forest Service and the Bureau of Land Management have long done everything in their power to subvert the intention of preservation legislation in all its forms. They have illegally manipulated the process of inventorying land for wilderness study in order to eliminate areas of conflict with timber, mining, and energy interests. They have allowed roads to be built into areas under consideration (and interim protection) for wilderness designation. They have issued oil and gas leases in critical habitats adjacent to national parks—as well as within areas under consideration (and interim protection) for wilderness designation. They have arbitrarily withdrawn wilderness study areas without public hearings when private interests have indicated a desire to develop the resources contained therein and have attempted to justify their actions with specious excuses and cockeyed record keeping.

One need look no farther than the Book Cliff/Desolation Canyon wilderness for confirmation. When the BLM completed its wilderness review of the area in 1980 (as mandated by the Federal Land Policy and Management Act of 1976, or FLPMA), it could only identify 362,000 acres (out of 15 million acres total) in five wilderness study areas (WSAs) that it felt might qualify for inclusion in the National Wilderness Preservation System (i.e., land possessing "opportunities for solitude or for primitive and unconfined recreation" and/or containing unusual "geological, ecological, scientific, educational, scenic or historical" values). Ordered by the U.S. Department of the Interior Board of Land Appeals to try again (92 percent of the sites it had thrown out were judged to have been improperly omitted), it managed to add 180,000 acres in the Book Cliffs area, found ways to delete acreage in other areas, and came up with even less. Quite simply, the

BLM recommended against WSA status for virtually every patch, plot, and divot with a potentially developable mineral reserve. And even when it did favorably recommend an area, as in the Book Cliffs, it violated FLPMA law by permitting oil and gas companies to perform 14 miles of seismic drilling and to punch seventeen exploratory wells inside its own WSA boundaries.

To conservationists, the whole BLM process of inventorying wilderness and recommending areas to be permanently protected was a joke, an in-your-face statement of contrary, sagebrush rebellion values. The Utah Wilderness Coalition (which included all the major conservation organizations, including the Sierra Club) recommended 718,600 acres be set aside in the Book Cliffs/Desolation Canyon area. The final BLM proposal, if it ever gets made, is expected to be 328,000 acres. Not a very funny joke. Pretty divergent values.

The town of Moab is about 40 miles south of Ray's Tavern—after another stage-set promenade (with the snowcapped La Sal and Abajo Mountains for backdrop) across plains of blackbrush and narrow-leaved yucca toward distant ramparts of salmon-colored mesas, down through the narrow canyon that separates Arches National Park from Canyonlands National Park, and across the bridge over the Colorado River into . . . Moab? Don't recognize this. Been AWOL a few years here. Maybe I never really noticed it before.

It seems to have multiplied since . . . well, okay, so maybe it's been twenty-five years. Who's counting? So I don't remember all those T-shirt and gift shops, Western wear stores, antique and collectible galleries, Indian crafts, artisan co-ops, delis/bakeries, bookstores, ORV rentals, bike rentals, rafting companies, helicopter tours, realtors. I count seven realtors, or is it seventeen? There must be three dozen places to eat and at least as many places to stay, though every one I cruise by has a "no vacancy" sign prominently displayed. I will learn that Hollywood is in town (with Robert Duvall, a 13-ton off-road camera crane, and God knows how many supernumeraries) making a $40 million movie about Geronimo, who never in his whole life came anywhere near Utah.

The Moab of memory offered five or six motels and your choice of two eggs any style, patty melt, prime rib. Now you can start your day with an olive pesto cream cheese bagel and a double decaf latte; lunch on penne pasta with shitake mushrooms; perk up mid-afternoon with a biscotti and a cappuccino (organically grown beans, of course); dine on eggplant parmigiana, Szechuan stir-fry, or, if your taste runs south of the border, pigs in a poncho. And you don't have to listen to Conway and Loretta while you eat. Moab has public radio on 89.7 FM, twenty-four hours a day.

In other words, it would appear to this befuddled old fart that Moab got "discovered" and now sports a growing in-migrant culture of nouveau buckeroos and buckerettes, and a transient population of sweat freaks and sprocket-heads in Air Mowabb cross-trainers and Cavrianas, who like to get out there in the rocks and stumps and bumps and dirt for a day or a lifetime get-away. Moab itself is still an admirably ugly town of junked cars and immobile homes, and its occasional attempts at gentrification seem almost satirical, but the country all around Moab is as stunning and dra-matic as you can find anywhere in the world, a great, silent land of brooding stone, secluded, private, wide open, mostly unregulated, and certified USDA choice by *Outside* magazine, travel advisor to the trendy. Add all the RVs, ORVs, ATVs, foreign visitors on world excursions, and the plain old moms and pops "spend-ing our kids' inheritance," and you have a space prob-lem. The question of whether canyon country can survive the admiration becomes more than academic.

Consider a few statistics. In 1982 the number of visi-tors recorded at Arches National Park was 339,415; in 2005 it was 781,670—over a 100 percent increase. The number of tourists to Canyonlands National Park in 1982 was 98,310; by 2005 it had quadrupled to 393,831. During the same period Bryce Canyon National Park's annual visitations increased from 704,796 to a whop-ping 1,477,324; Zion's from 1,361,750 to 2,687,848; and Capitol Reef's from 323,458 to 729,324. And the same pattern is true for Natural Bridges, Cedar Breaks, Rainbow Bridge, and Glen Canyon National Recre-ation Area. Where do all these sightseers come from? According to Jay Woolley, former director of the Utah Travel Council, 26 percent listed Southern California as home. But of course. Has anyone tried visiting a park in California lately?

There are a lot of Moab residents who aren't so enthusiastic about the snowballing penchant for uncontrolled recreation (not to mention relocation) in their own backyard. And understandably. My friend Bill Hedden, a longtime resident of Castle Valley, once told me of an afternoon one spring when he was coming back from Paiute County along Highway 128 that parallels the Colorado River. "I suppose I've gotten used to the increasing volume of traffic, but this time things really got my attention. I came around a bend a few miles from home and there was some guy standing up on a bluff driving golf balls into the river. There were campers jammed into every conceivable site where you could pull off along the banks, jet skiers were out there in the current competing with rafters, hot air balloonists competing with helicopter flights over the canyon; it was nuts. We are simply going to have to find a way to regulate recreational usage or the entire area is going to be completely destroyed—what's left of it."

Consider another ghastly statistic. In 1982 the number of mountain bikes in the United States was 250,000, a good many of them out there tearing up the nation's backcountry even then. In 1994 there were 25 million. God only knows how many there are today. And ever since *Outside* magazine honored Moab as its "favorite mountain-bike spot," it has been completely overrun by bikers attracted by the "gonzo/abusive" slickrock trails. In 1989 some 26,000 of them rode the Slick

Rock Trail up above the Moab dump. A year later, 50,000. In 1992 the numbers nearly doubled again. According to the *High Country News*, 1.2 million people used Moab-area BLM lands in 1996, 170,000 alone at the Mecca of mountain biking, Sand Flats and the Slick Rock Trail.

It may seem to the uninformed that this non-motorized, non-polluting use of the back country is more benign than metal-mashing jeep safaris, drilling rigs, and movie companies with 13-ton boom cranes. But it is not. Damage to fragile ecosystems by the gonzo/abusive has been appalling. As Ken Davey, feature writer for the *Canyon Country Zephyr*, observed in one of his columns, "Dozens of square miles near the Slick Rock Trail along Sand Flats Road have been destroyed, and they will stay that way for generations. On any weekend during the spring, choking smoke fills the canyon from campfires along the River Road. Portable pit toilets along the river corridor haven't solved the problem of human waste." An article in *High Country News* in November 1991 points to the stench of human excrement and the accumulation of garbage along the Colorado River and quotes Southern Utah Wilderness Alliance attorney Scott Groene as saying "At first people thought, yeah, great, tourism, it's great, it's clean. But now people are running around out of control, looking for a place to camp, pushing into sensitive areas. The desert bighorn barely survived the uranium boom; now the mountain bikers may finish them off."

Storm troopers of the fat tire battalion may finish

off more than the wildlife. Consider the cyanobacte-
rial crust (also called cryptogamic crust) that accounts
for three-quarters of the living ground cover on the
Colorado Plateau. It's that black stuff you're tramp-
ing through over there that looks like a lumpy lawn
of dead moss, but it's not dead and it may have taken
seventy-five years to establish itself in the area you just
trashed. Cyanobacteria occurs in the desert most commonly as filaments, filaments surrounded by mucilagi-
nous sheaths that when wet push through the soil to
create a complex network of fibers (*Microcoleus vagi-
natus*) and hold it all together, impervious to erosion
by wind and water. Cyanobacteria swells to ten times
its size when saturated by rain, storing essential water
supplies for vascular plants and crustal organisms, and
it also contributes critical amounts of nitrogen into
the desert ecosystem. Crush it with a boot or a hoof
and it's history. Crush it with a tire, and you hasten the
process of desertification by adding a continuous track
that is even more susceptible to erosion.

Perhaps the invasion of Moab would be of little
more than local concern if it were not symptomatic
of a pattern occurring all over the Colorado Plateau,
and it has for some time now become manifestly
clear to people concerned with protecting the environment that unregulated tourism is as big, if not a
bigger, threat to the ecological integrity of land than
range mismanagement, mining, nuclear waste, coal-
fired power generation, and deficit timber sales.

* * * * *

Well, the storm troopers have certainly finished me off. I'm outta here, shaking a palsied finger at a passing convoy of ORVs and mumbling into my whiskers. Going to sneak up to Murphy Point and say one last goodbye to this once-upon-a-time outpost of Eden. Murphy Point is out along a washboarded, four-wheel-drive track that wanders off through piñon/juniper woodland from the paved road that traverses the Island in the Sky, that great mesa that begins somewhere in the flats above Tenmile Canyon and ends in a precipitous drop to the confluence of the Green and Colorado Rivers at the mouth of Cataract Canyon. A thousand feet below the broad bench of the White Rim, a 200-foot-thick layer of Cutler Formation sandstone left by the final retreat of a Permian sea is already half in shadow. A few miles to the west and another 1,000 feet below the White Rim, I can see into the dark reaches of Stillwater Canyon and just a glimpse of the Green River, just a fingernail where the fading light catches it at the bend of a meander. Fifteen miles to the west, beyond the river and across the northern extension of the Glen Canyon National Recreation Area, the Orange Cliffs rise in an undifferentiated wall at the edge of the San Rafael Desert, and in the far distance (now we're talking 80 to 100 miles), Boulder Mountain and the Aquarius Plateau block my view into Nevada. Well, Nevada is 300 miles away; maybe it's not quite that clear this evening.

Like almost any promontory along the rim of this

mesa (Dead Horse Point and Grand View being the most visited), Murphy Point overlooks a 280-million-year-old layer cake of inconceivable, preposterous, sedimentary rocks. A geologic hallucination. Every scribbler before me has apologized for his inability to communicate the lyricism of the view—any view—out here on the plateau, and I hasten to join the ranks. I don't care how many inspired images of gothic architecture and sacred monuments I invoke, I'm still not going to come close. Forget it. It's just a pile of rocks. It's just beachfront laid down by the to-ing and fro-ing of intermittent seas, by erosion from the Uncompahgre Uplift to the north, by blowing sand dunes that covered most of Utah during the Triassic and Jurassic periods. It's just detritus left in the wake of vast erosional forces that completely stripped 140 million years of deposited formations off the top (all that Dakota sandstone and Mancos shale—a vertical mile of everything, in fact, that once lay above the Navajo sandstone) and went on cutting down through the Kayenta, Wingate, Chinle, Moenkopi, Cutler, and Honaker formations to the ancestral rock of the Paradox formation revealed at the bottom of Cataract Canyon.

Or, as one Mormon rancher trying to make a living in the midst of all that scenery once said, "It's a hell of a place to lose a cow." A frustrated resident of Garfield County once offered my father and me this opinion, as we argued one afternoon in the town of Escalante over the proposed paving of the Burr Trail: "Red ledges!" he said. "Red ledges! What's so goddamn exciting about red ledges?"

The wind comes at me like an attitude across those red ledges where I have assumed my twilight perch. It seems chipper and playfully enough disposed, though at the same time mercurial and wicked, as if possessed of a demoniac presence. I'm not much given to personification; I would have made a lousy Indian. But this wind, I think, flirts with me; I hear it coming like black riffles approaching across water, then flitting suddenly away to boogie in the rice grass and yucca. Without so much as a kiss, the harlot. I snub its next advance and it grabs my Charlie Tweedle 4X imitation beaver cowboy hat and flings it into the Jurassic void.

Sitting here in near absolute silence, looking out over hundreds of square miles of incomprehensible grandness, I think, and not altogether whimsically, that I have in some measure my father to thank for this view. In 1961 he came here as a special assistant to Secretary of the Interior Stewart Udall, and together with Joe Carithers, Dean Brimhall, Bill Kruger, and Bates Wilson spent days trekking through the South Desert, Cathedral Valley, and the Waterpocket Fold, looking for areas that might be included in an enlarged Capitol Reef National Park (at that time a national monument). It was Bates Wilson, superintendent at Arches National Monument, who promoted the idea of Canyonlands National Park (and eventually became its first superintendent), but it took the foresight of Udall and all those who worked with him to transform vision into reality.

I say we lock the sonofabitch up and shoot trespassers—which means everybody. Or at the very least we institute a permit system for backcountry use on *all* public lands, periodically retiring for, say, three to five years (or maybe thirty to fifty years) any that begin to show signs of serious wear. If we can put people on a list and make them wait nine or ten years for a private permit to run the Colorado River through the Grand Canyon, we can make the same rules for a jeep safari, or a gonzo/abusive bike ride on Wilson Mesa, or an hour at the Colorado River driving range along Highway 128. Recreation, like livestock, mining, and timber, is a federally subsidized activity on the public lands, and there is no reason it should escape regulation. Anyway, what's so abhorrent about "locking up" a few more red ledges? We need to take seriously the proposition that the idea of wilderness is more important to the human spirit than any "use" (occupational or recreational) we put it to. To quote my old man, the plateau country is "not a country of big returns, but a country of spiritual healing, incomparable for contemplation, meditation, solitude, quiet, awe, peace of mind and body." We can create industrial and social nightmares anywhere and anytime we feel like it—unless we think we've finally got enough. But "wilderness, once we have given it up, is beyond our reconstruction."

THIS IS DINOSAUR?

or

"Dude, Where's My Park?"

I HAVE ELSEWHERE confessed my inability to retain for more than two paragraphs any of the geological data I have serially attempted to memorize over the years—all those Precambrian, Cambrian, Triassic, Jurassic, Carboniferous whatyacallems, those indistinguishable shales, sandstones, limestones, siltstones, mudstones, Rolling Stones that always show up in a cheerfully colored stratigraphic column in the "story behind the scenery" booklets I always buy at the visitor center. But on Highway 40 from Vernal, Utah, to the Yampa River launch site just over the Colorado line, you can see for yourself all that relevant stuff in living color—the Mancos shale, the structural downwarp of the Lily Park syncline, the grassy terrace of Deerlodge Park where the Yampa suddenly disappears behind a monoclinal bend in the Weber sandstone and vanishes into the deep Miocene canyons of Dinosaur National Monument.

As the kid driving our shuttle observes, the visuals are totally awesome, a real jeez-Louise, wait-till-the-folks-back-home-see-this photo-op of dun-colored hills, benches, hogback ridges, buttes, mesas, arroyos, the whole business back-dropped by the draggle-assed eastern end of the Uinta Mountains, the middle ground peppered with clumps of gray-hued sage, greasewood, and rabbitbrush, and right down there we have a long green slash of cottonwoods and coyote willow marking the river corridor and providing a wee bit of color.

From its headwaters some 200 miles to the east in the Colorado Rockies, in a pond- and lake-dotted wilderness area called the Flat Tops, the Yampa flows north to Steamboat Springs, then turns due west across the plateau country behind the Williams Fork Mountains and the Danforth Hills. It picks up the Elk River near the town of Milner and the Little Snake just before reaching Deerlodge Park, then bores directly into the steep, towering canyons of Dinosaur National Monument. Descending at an average rate of 11 feet per mile, it terminates at its confluence with the Green River in Echo Park, smack dab in the middle of the monument, and some 46 miles and 533 feet below Deerlodge (where there is no lodge, by the way, unless you happen to be a deer).

But what a magnificent 46-mile descent it is. True, the Yampa is a geezer's delight, mild, gentle, benign, hospitable, without a rapid worthy of the designation except Warm Springs. But having personally reached the age of geezerdom at the time of this adventure (May 2001) I am more than happy to scooch back against the duffel behind my rowing seat, prop the oars under my knees, and just let the raft float lazily with the current.

Sometimes I see where I'm going, sometimes where I've been, sometimes nothing but the shadow of floaters on my retina, which the ophthalmologists describe as a breakdown of the vitreous humor, though I don't see what's funny about it.

I observe no tongues of slick-water leading into channels choked with rocks, no sebaceous downriver humps indicating submerged boulders hiding boat-eating holes, no snags, sandbars, keepers, bottom rippers, forever eddys (or at least none that I remember). As David Bradley remarked in his chapter "A Short Look at Eden" in *This Is Dinosaur: Echo Park Country and Its Magic Rivers*, "The Yampa is of the desert, a long Canyon de Chelly blessed with the magic carpet of moving water."

Indeed, my greatest excitement, other than musing about what's for dinner, is the paperback edition of Powell's *Exploration of the Colorado River and Its Canyons* that I keep in the ammo can beneath my seat and dip into now and then when I get tired of ogling canyon walls and interspersed benches of cottonwood and box elder. I also have in the can *This Is Dinosaur*, a book edited by my father in 1955 as part of the explosive opposition to a Bureau of Reclamation proposal to build two impoundment and hydroelectric dams within the confines of the national monument, one at Echo Park and the other at Split Mountain. The fight to protect the integrity of the park system itself (amendments to the Federal Water Power Act in 1921 and 1935 made national parks and monuments off-limits to water projects) and Dinosaur in particular (the reservoir behind these dams would have extended 63 miles up the Green River and 44 miles up the Yampa, essentially drowning the entire refuge) was the first major battle between conservationists and the bureau since John Muir and his fledgling Sierra Club tried to prevent the flooding of Yosemite's Hetch Hetchy Valley by the City of San Francisco in 1913. Unfortunately, Muir lost. But the Dinosaur preservationists won, and their struggle, as a number of historians have noted, was a major contraction in the birthing of the modern environmental movement.

It also considerably facilitated our trip down the river and left unimpaired for future generations a dry campsite called Ponderosa at river mile 36.8. A colluvial terrace, Ponderosa, with a relic pine tree of that particular species growing out of the cobble almost at water's edge and a wall of Round Valley limestone on the agenda downstream. On the *dinner* agenda, smoked oysters and crackers, marinated steaks, oven-roasted potatoes, Caesar salad, foil medley of vegetables, pineapple upside-down cake.

I am not here all on my lonesome, needless to say. Most of the usual suspects are sedately positioned in their respective water crafts, eager, as I am, for the solitude of a few days on a wilderness river. But I have a secondary motive, which is to see for myself these canyons that my old man played such an instrumental role in saving from the dam builders and that he never actually witnessed for himself. Somebody in the family has to do it.

Tepee Rapid, a half mile downriver from Ponderosa, provides a few hundred yards of splashing next morning but otherwise fails to get the heart pumping. Neither does Little Joe Rapid, Five Springs Rapid, or Big Joe Rapid, all of which we *ooh, ahh* through during the 17 miles before camp two at Harding Hole. No doubt it is irresponsible to suggest that the Yampa is in no respect hazardous—any river can be hazardous, and there have been several drownings at Warm Springs Rapid since its formation in 1965—but with today's inflatable boat designs and type-III personal flotation devices (PFDs) that will hold one's head out of the water even after it has been whacked out of line on a rock (assuming, that is, that the PFD is being *worn* at the time), not to mention all the throw bags, ropes, pulleys, carabiners, and other rescue paraphernalia now routinely carried in river rafts, one would have to be pretty careless or phenomenally unlucky to wind up feeding the squawfish and chubs on the Yampa.

Of course there *was* that infamous party of four who were sent out in August 1928 to "explore" the Yampa and photograph its wonders for the *Denver Post*. They had two wooden boats, appropriately named the *Leakin Lena* and *Prickly Heat*, both 16.5 feet long, 4.5 feet wide, and during their proposed two-week trip these stalwarts managed to run out of food; lose their camp gear, two cameras, and most of their film; and turn the *Leakin Lena* into kindling against some rocks, all during the first 15 miles of their adventure. The *Post* headlines read "Two Barely Escape: Pair Pitched from Boat in Yampa Crash," and later "Death Faced Many Times by *Post*'s Quartet of Explorers," and finally "Glad to Get Home Alive! *Post*'s Yampa Canyon Expedition Safe." Safe because they abandoned the whole shiteree at about mile 12 1/2 and fled the river up Hell's Canyon in the direction of civilization. Nobody died on the *Post* expedition, though the following year there were four drownings, all apparently due to inexperience and experimental boats.

During May and early June 1965, heavy rain fell almost continuously over the eastern Uintas, seventeen out of twenty or twenty-one days, according to historian Roderick Nash, saturating the slopes above the Warm Springs drainage and turning them into a kind of precariously perched pudding. On June 10, those far up Warm Springs Draw began to slide, picking up mud and scree as they descended from the steeper canyons, eventually turning the whole gathering slurry into a 15-foot wall of accelerating rock and gravel that must have hit the Yampa like a freight train. It completely dammed the river for a time, before the current ate its way through and began washing the smaller stuff downstream toward the Green. What was left behind was a boulder-strewn channel a half-mile long and crooked as a snake.

The following morning, two 27-foot pontoon rigs belonging to Hatch River Expeditions came floating peacefully down with a party of Boy Scouts from

Denver, when to their utter surprise, they suddenly came upon a violent, churning maelstrom with waves approaching 20 feet that the experienced boatmen knew could not be there. But it was there. Too late to pull over and scout, they were forced to run it blind and, at least in the case of the lead boatman, Les Oldham, without a life jacket. In the enormous hole center-right at the bottom of the entry tongue, Oldham's oarlock snapped and he was ejected backward out of the boat, which was the last seen of him until his body washed up in Island Park seventeen days later.

Mindful of this sad misfortune, I spend extra time during our scouting stop at Warm Springs staring at the 10-foot waves at the end of the tongue, waves that try to force one into that seriously unpleasant hole where the Hatch boatman went overboard. Need a downstream ferry angle to the right to punch through those waves, and I'd better time it to be in the trough, or I'm going to catch enough crabs to make a Louie. This totally commands my attention and dries up my mouth so bad I feel as if elephants have been dusting themselves in there. And in my distracted state I more or less forget to look at the rest of the rapid—which turns out to be a mistake, because there are a number of huge midstream rocks that must be avoided, a half-dozen somber pourovers and holes all the way down, and a boulder bar to rock and roll on. All of this it seems I manage to hit (head-on, fortunately), with eye-popping, denture-jarring, whiplashing élan—avoiding a flip or a wrap (*quel elegance*)—but not deafened by my colleagues' applause. I concede that perhaps there is a reason Rod Nash (*Wilderness and the American Mind*) devotes a chapter to Warm Springs in his book *The Big Drops: Ten Legendary Rapids of the American West*. Still, I don't think I'd put it in the same league with some of the other legends—Lava, Crystal, Granite, Satan's Gut, or, for that matter, Snaggletooth, which didn't make the cut.

Four miles downstream from Warm Springs, the east-west-flowing Yampa ends in a marriage with the north-south-flowing Green River at Echo Park, in the very heart of the monument. Dead ahead, as one comes bobbing out into the confluence, there looms a colossal wall of Weber sandstone called Steamboat Rock. More than 1 mile long and 1,000 feet high, it forms the core of an entrenched meander and is nearly circumnavigated by the Green as it makes a long hairpin turn from south to north before angling off again to the west. John Wesley Powell, the first serious explorer of these rivers in 1869, described this wannabe Rincon as a "peninsular precipice with a mural escarpment along its whole course on the east, but broken down at places on the west." He called it Echo Rock, because the voices of his men standing beneath the cliff walls on the opposite side of the river from Steamboat Rock were repeated back "with star-tling clearness, but in a soft, mellow tone, that trans-forms them into magical music." So infatuated with the melodic metaphor was he that he named the wall behind him Jenny Lind Rock, after the famous nineteenth-century opera soprano otherwise known as the "Swedish Nightingale."

Steamboat, née Echo Rock, proved too great a temptation for Powell to just stand there listening to it reverberate. Attempting to ascend it, he and one of his men, George Bradley, clambered over piles of broken rock and a series of benches until they reached a height above the river of some 600 to 800 feet—whereupon they encountered a near vertical wall just below the rim. They found a place where it seemed possible to continue upward, and, as the one-armed major described it, "We proceed stage by stage until we are nearly to the summit. Here, by making a spring, I gain a foothold in a little crevice, and grasp an angle of the rock overhead." He now finds himself stuck. Perhaps he has forgotten about his missing arm. Whatever—on this near perpendicular pitch he is not only unable to let go and grab another handhold, but also he is unable to step back and reach his original foothold without seriously risking a backflip onto the rocks some 80 feet below.

Bradley climbs above and looks around for something to poke over the ledge that the major might grab onto. Nothing doing. Hanging there by his toes and his remaining five fingers, Powell's muscles begin to tremble and shake, and he is no doubt conjuring up the image of himself as a corpse, when it suddenly occurs to Bradley to remove his pants and dangle the legs down near Powell's trembling hand. Powell makes a grab, Bradley hauls away, and the catastrophe is averted. "Then," says Powell, with characteristic passion, "we walk out on the peninsular rock, make the necessary observations for determining its altitude above camp, and return, finding an easy way down."

Exhausting even to read about all these goings-on. I think another baloney and mustard sandwich is in order, followed by a nap under this magnificent grove of box elders above the beach here at camp three, aptly named Box Elder.

No such slothful inclinations beset the major, I must say. The day after he nearly did his head-plant off Steamboat Rock, he had his men row him 5 miles *up* the Yampa, swollen with spring runoff, and then climbed several thousand feet out of the canyon, just like the bear who went over the mountain—and for the same reason. What he saw, among other things, was a considerable portion of the Yampa canyon to the east and a number of peaks on the northern rim, the highest of which he named Mt. Dawes. And, of course, having seen and named, he had to return the following day and climb. From the summit of Mt. Dawes he reported, "I can look away to the north and see in the dim distance the Sweetwater and Wind River mountains, more than 100 miles away. To the northwest the Wasatch Mountains are in view, and peaks of the Uinta. To the east I can see the western slopes of the Rocky Mountains, more than 150 miles distant."

I will take the major's word for it. My sentiments echo those my father once defined in his famous argument for the concept of wilderness ("The Wilderness

Letter"): that those who are too old, too infirm, or
too shiftless to actually *go* to wild country "can simply contemplate the *idea*, take pleasure in the fact that
such a timeless and uncontrolled part of earth is still
there." I do.

It's a short ride from Echo Park to the mouth of
Whirlpool Canyon, with its high, sheer cliffs dropping straight into the narrow channel of the now
combined Green and Yampa Rivers. It is here that
the Bureau of Reclamation proposed in July 1947 to
build the first of two dams as part of its water resource
development plan for the upper basin participants in
the Colorado River Compact of 1922. Under the terms
of that compact, Wyoming, Colorado, Utah, and New
Mexico were guaranteed 7.5 million acre-feet of the
Colorado's annual runoff, but the specific allocation of
that flow (which, by the by, was grossly overestimated)
was left for the states themselves to decide—a wran-
gling debate that took some twenty-five years. No reclamation projects could be authorized until the Upper
Colorado River Commission came to an agreement
on who was to get what, but once it did the bureau
would be good to go. Or so it thought.

While it waited, both Echo Park and Split Mountain
were extensively investigated as potential dam sites—
Echo Park eventually being judged clearly superior.
But just like its counterpart in nature, the beaver, the
bureau had never been able to stand the sight of running water and, faced with making a choice, decided it
would build them both. Accordingly, Echo Park would

be a concrete gravity dam, 529 feet high and capable of generating 200,000 kilowatts of power; Split Mountain would be much smaller and designed primarily to impound peak releases from Echo Park. The storage capacity behind these plugs was estimated to be 6.4 million acre-feet and, as earlier noted, would create a recreational septic tank that would extend some 63 miles up the Green River's canyons and 44 miles up the Yampa's. "Tomorrow's Playground for Millions of Americans," one pamphleteer put it, visions of speedboats dancing in his head.

Not everybody was so excited, among them Bernard DeVoto. DeVoto attracted national attention to the dam-builder's scheme with an article in the *Saturday Evening Post* (circulation 4 million) in which he railed against development within any national park or monument boundary, pointing out that such a thing was a violation of the "use without impairment" clause contained in the National Parks Act of 1916, and he blasted away at the aspiring defilers of Dinosaur Monument in particular. "If it is able to force the Echo Park project through," he sarcastically observed, "the Bureau of Reclamation will build some fine highways along the reservoirs. Anyone who travels the 2000 miles from New York City—or 1200 from Galveston or 1000 from Seattle—will no doubt enjoy driving along these roads. He can also do still-water fishing where, before the bureau took benevolent thought of him, he could only do white-water fishing, and he can go boating or sailing on the reservoirs that have obliterated the scenery." Pointing out that these pleasures were all available to New Yorkers and Galvestonians without having to travel anywhere, he concluded, "The only reason why anyone would ever go to Dinosaur National Monument is to see what the Bureau of Reclamation proposes to destroy."

The editor of the *Denver Post* took umbrage and wrote an editorial not only condemning DeVoto's article but opining that Westerners could do what they wanted with their scenery. He was getting in over his head. In a reply dated August 1, 1950, DeVoto sliced and diced: "As a native Westerner from Roseville, Illinois, do you read the texts you quote, misunderstand them, or merely misrepresent them? The point of my *Saturday Evening Post* article was entirely different from what you tell your readers it is. My point was that all the values which the Bureau of Reclamation alleges will be derived from the still illegal Split Mountain and Echo Park dams could be secured by dams built outside Dinosaur National Monument. Now that I have explained what you were supposed to have read, please inform your readers."

There followed several paragraphs further excoriating the *Post*'s reportage and its editorial practices, and then this: "You are certainly right when you say that 'us natives' can do what you like with your scenery. But the National Parks and Monuments happen not to be your scenery. They are our scenery. They do not belong to Colorado or to the West, they belong to the people of the United States, including the miserable unfortunates who have to live east of the Allegheny hillocks."

The *Reader's Digest* reprinted the *Saturday Evening Post* article, further publicizing the issue, and as public opinion against the dams increased over the next few years, the entire Colorado River Storage Project began to look in jeopardy. Five years after DeVoto's article, and with a citizenry aroused by the continued efforts of a united conservation movement (led by Dave Brower and the Sierra Club), Congress passed a compromised Upper Colorado River Storage Act that deleted Echo Park and further stated that "It is the intention of Congress that no dam or reservoir constructed under the authorization of this act shall be within any national park or monument." God was back on his throne; all was right with the world. Well . . . sort of. The compromise *behind* the compromise was that the enviros would not oppose the construction of a dam on the Colorado River in Glen Canyon, or anywhere else outside of national park and monument boundaries—a capitulation that in time many would come to believe was the worst mistake they had ever made.

The increase in water volume from the convergence of the Green and Yampa creates great whirlpools and eddies as the current caroms off the sheer rock walls in Whirlpool Canyon, and once again it's tuck in the oars and pinball down to Island Park, where everything suddenly opens up, and the current slows to a lazy shuffle around a number of small islands and bends and along low banks revealing an alarming population of mule deer. It's an expansive canvas of cotton-woods and grassy vistas speckled here and there with prickly pear, sage, juniper, rabbitbrush, and flowers—larkspur, penstemon, paintbrush—all fading back into the low foothills that bound the broadly eroded Island Park syncline to the north, west, and south. Diamond Mountain Plateau forms the horizon on the right; the Blue Mountain anticline rises steeply on the left.

At the end of this big hairpin meander, with its multiple channels, we come into Rainbow Park and at its far end the maw of Split Mountain Canyon—the approach to which reminds me, for some reason, of the great grinning chops of a baleen whale. Maybe it's that the folded rock arching high above where the river suddenly disappears looks like a breaching humpback—or maybe it's just I've had too many cans of preprandial suds.

The river through Split Mountain drops more rap-idly than in any other part of the monument, an aver-age fall of some 19.3 feet per mile—or 146.68 feet over its total length of about 7.5 miles—which means the runs through Moonshine, SOB, and Schoolboy Rap-ids are fast and demand a boatman's semi-attention. The whole canyon goes by very quickly. Blink and you miss it—but not that pourover dead ahead or the hole and reversal behind it.

And then, all of a sudden, it's over. Here we are at the take-out at Cottonwood Wash. Who are all those people running around, all those parked cars, all those poop tents pupped in the Split Mountain campground?

It's depressing after four or five days of relative isolation to be suddenly back in the workforce, lugging rocket cans, frames, oars, coolers, kitchen boxes, and duffel bags up to the vehicles in the furnace heat of the Uinta Basin. As a final gesture of bonhomie and just to perk myself up, I volunteer my sweet wife Lynn to dump and clean the port-a-pot at the waste disposal site discreetly positioned away from the boat ramp and parking lot. Rest assured, she'll have no trouble finding it. They couldn't hide it from her if they'd put it in West Virginia.

Postscript

On our way out of the monument, Lynn and I stop at the dinosaur quarry visitor center a few miles up from the river. The quarry site, discovered in 1909 by paleontologist Earl Douglas, contains the greatest deposit of dinosaur bones in the world, which is the draw for 95 percent of the people who come to this out-of-the-way location in the first place. The visitor center is a spacious building, modern in design, built right up to and enclosing a long section of an ancient riverbed tilted during the Uinta Mountain uplift into a nearly vertical hogback ridge of Morrison formation sandstone some 200 feet long. Over time, eons of time, 5,000 feet of erosion exposed in its thick sedimentary layers a great, stacked burial ground of dinosaur bones. More than 2,000 of them are fossilized in that one wall beneath the quarry's sheltering structure, where the air conditioning is purring nicely, thank you, and cold drinks are on sale in the museum shop. This is not just some "see the terrible lizards" tourist trap along the desert highway between Vernal and Craig; it is truly one of the most phenomenal prehistoric exhibits in the world, and, indeed, it attracts people from all over the world. It is the reason Woodrow Wilson established the monument in the first place, and a good part of the reason Franklin D. Roosevelt expanded it in 1938 to cover an additional 209,000 acres. (Bryce Canyon is 35,835 acres; Arches, 73,234; Zion, 147,035. Only Canyonlands is bigger, at 337,587 acres.) It is only Dinosaur's status as a monument that suggests, inaccurately, that it is a lesser jewel in the West's crown of national parks.

Post-postscript,
this time six years later, in May 2007

Lynn and I are returning from Salt Lake to Santa Fe and decide to take the long way around and revisit Dinosaur. Diving up the road to the visitor center at the quarry, we find the road closed, yellow crime tape barring the way, nobody in the kiosk where I try to present my Golden Geezer Pass, no snappy-looking ranger in her persimmon pudding sombrero to hand us welcoming maps, informational brochures, and lists of all the things we are not permitted to do. There's nothing but a beshat concrete replica of some hangdog, saurian reptile and a single-wide house trailer with a sign on it saying DINO-STORE, where I purchased my "story behind the scenery" pamphlet and a purple plastic *Stegosaurus* (no relation) for my grandchildren.

"Dude, where's my park?" I inquire of the attendant.

"There's a man in the temporary exhibit facility who can give you the story behind the non-scenery," he replies (or something like that).

We find the temporary edifice that we had somehow overlooked and come across a very pleasant, if somewhat lonely, park service volunteer soldiering on in an empty room with the femur of something very large on display and a relief map of the monument and a few pictures of whiskery gentlemen posing as geologists. (I invoke poetic license here; I actually don't remember anything about this space except the very amiable man.) "Dude," I say. "Where's my park?" (or something to that effect).

"Closed," he says apologetically, "but if you want to take the fossil discovery trail, you can see some leg bones and a vertebra. It's only about a mile round-trip."

And then he explains that what we really came to see, that architecturally marvelous visitor center built in the 1950s by Anshen and Allen and declared a National Historic Landmark in 2001, housing its great graveyard of 2,000 fossil bones (and you're trying to fob off a *leg* and a *vertebra*?), had gradually begun to slide down the slope on which it was built until eventually a part of the roof fell in and they had to close it.

Why weren't measures taken to stabilize it? Good Lord, why wasn't it repaired?

"Well," says our rueful aide, "funding, I guess."

He is too polite to say, "For the same reason you're talking to *me* and not a paid employee of the National Park Service. Go ask the cheap, chiseling, public-lands-hating, privatize-the-national-parks Bush administration."

The truth is that the center, magnificent though it may have been, was constructed during the "build in haste, repair at leisure" era in the 1950s and has had problems from the very beginning. Cracks in the parking lot appeared before it was even open; an unsettling vibration in the upper gallery required installation of some major columns; there were problems with drainage, buckling floors, windows cracking under stress, and what was referred to in various structural reports conducted over the past forty years as "adverse movement." Basically the building was sliding around because it was built on layers of sandstone and bentonite shale, and bentonite is moisture-sensitive. It loses its structural integrity when wet. As one engineer put it, this was "a magnificent structure built on sand."

Although one can't dump these long-term problems on the current Department of the Interior, the Bush administration has made no effort to address the present predicament—other than to close up shop. The administration's mugging of the entire planet has been mind-boggling, to say the least, much of it out of stupidity, indifference, and neglect, but in its passion to open the nation's parks to development and to turn their management over to private concessionaires, it has been relentlessly and deliberately focused. As the former superintendent of Yellowstone, Mike Finley, said in a speech back in May 2003, "I worked for four Republican Presidents and two Democrats and during the course of that career, never have I witnessed such an ideological war on the natural resource laws, policies or practices or institutions. Even our national parks are not safe from this assault."

The easiest way to conduct this onslaught is to divert or withhold funding for maintenance, repair, and operations. A February 20, 2004, memo sent to park superintendents in the Northeast Region by Chrysandra Walter, deputy director for the Park Service's Northeast Region division, suggested the following frugalities:

> Close the visitor center on all federal holidays.
> Eliminate life guard services at 1 of the park's 3 guarded beaches.
> Eliminate all guided ranger tours.
> Let the manicured grasslands grow all summer.
> Turn 1 of our 4 campgrounds over to a concession permitee.
> Close the park every Sunday and Monday.
> Close the visitor center for the months of November, January & February.

The memo went on to suggest that park personnel not allude to these strategies as "cuts"; rather, they should be referred to as "service level adjustments."

For Dinosaur National Monument, these adjustments have been like the second coming of the "Permian Triassic Extinction Event"—a somewhat upmarket turn of phrase to describe the asteroid collision that occurred 65.5 million years ago, altering the earth and doing away with all those Jurassic Park behemoths in the process.

But at least the interior of Dinosaur National Monument is safe, right? It's a wilderness, after all, its canyons isolate, inviolate and its rivers protected from impoundment. They can't be Bushwhacked. Or can they? The Northern Colorado Water Conservancy District would like to have a crack at it, as revealed in a 2003 article in the *Greeley Tribune*, "Study Shows Diverting Yampa River Could Help Colorado's Future Water Needs." (So would an outbreak of Ebola in Denver—a better solution, in my opinion, than the proposed rerouting of 300,000 acre-feet of the last free-flowing river in the West and delivering it through a series of pumps, pipes, and tunnels to lawns and swimming pools along the Front Range.) As David Brower said to John McPhee in *Encounters with the Archdruid*, "Our side never wins battles like these. We get no more than a stay of execution."

John Wesley Powell couldn't see 100 miles in any direction today. Air pollution from the oil and gas industry that has completely taken over the Uinta

Basin in northwestern Utah and all of Garfield, Rio Blanco, and Moffat Counties in northwestern Colorado has reduced visibility to half that distance, and as we head back through a kind of murky haze that leaves the watery-eyed wanderer wondering where's the forest fire spreading all the smoke, we wonder if the next time we'll decide to take the long way around. I am reminded of my father's words in his introduction to the dinosaur book, "We are the most dangerous species of life on the planet, and every other species, even the earth itself, has cause to fear our power to exterminate. But we are also the only species which, when it chooses to do so, will go to great effort to save what it might destroy." Which, I wonder, will we ultimately choose to be?

THE JEWEL
OF THE COLORADO

FOR MANY YEARS Glen Canyon was the solitary domain of a few Indians; a few prospectors; a few river runners like Norman Nevills, Art Greene, Ken Sleight, Al Quist, Georgie White, and Moki Mac; and a few wandering mystics like Edward Abbey, Ralph Newcomb, and Everett Ruess. It didn't gain much notoriety until it was too late to save it from the federal dam builders, who had finally agreed to abandon their designs on Echo Park in Dinosaur National Monument if the environmental groups who opposed them in that venue would leave them alone in Glen Canyon.

It was a Faustian bargain, and for nearly forty years after the annihilation of Glen Canyon its passing was lamented by dozens of writers and photographers, among them David Brower, Wallace Stegner, Edward Abbey, Gregory Crampton, Philip Hyde, and Eliot Porter. The consensus was that it had been lost to all mankind for all time; in Brower's words, "Nothing our technology will have taught us, in this century or any other, will be able to put Glen Canyon back together again." Abbey suggested, "To grasp the nature of the crime that was committed imagine the Taj Mahal or Chartres Cathedral buried in mud until only the spires remain visible. With this difference: those man-made celebrations of human aspiration could conceivably be reconstructed while Glen Canyon was a living thing, irreplaceable, which can never be recovered through any human agency."

Perhaps not. But none of these prophets foresaw the possibilities that might be brought on by a thick, juicy, well-marbled drought—say a ten or twenty year drought like the one that major parts of the West are currently somewhere in the middle of today, and one that global warming may turn into a permanent state of affairs. Folks who gag at the very thought of a restored Glen Canyon point cheerfully to the fact that this past winter was wetter than normal, but anyone who lives in the intermountain West or Southwest knows in which direction the winds are blowing.

In that context I received, in the spring of 2005, a call from Bruce Clotworthy, then working for an environmental organization called the Glen Canyon Institute in Salt Lake City, Utah. Lake Powell is actually drying up, Bruce said. Over the past five years its waters have fallen nearly 145 feet, and major features of the legendary canyon drowned beneath its noisome wavelets are starting to reappear. Cathedral in the Desert, the iconic symbol of all that was "lost to mankind forever," is once more a hike, not a scuba dive. Would I be up for a trip to go and see it?

Yes, I would. And so would my wife and daughter, and Philip Fradkin and his son, and Bruce's father, a former gold medalist at the Melbourne Olympics and a Peace Corps volunteer in Venezuela during my shaky tenure as one of its "deputy" directors. We all would. Certainly more of us than Bruce bargained for, but swallowing hard, he promises he will commandeer the institute's boat and put it all together. Meet him on the twenty-first of May at Lake Powell, "the jewel of the Colorado"—which is not a lake at all but a barren, power-plant reservoir almost totally devoid of vegetation or animal life that extends for 150 miles north into southern Utah from its impoundment site at Page, Arizona.

Traveling from our various home ports, we assemble on the appointed morning at Bullfrog Basin, an eyesore on the western shore near the upper end of what the tourist bureaus think of as a recreational paradise. Bullfrog and its cross-pond counterpart, Halls Crossing, are the primary access points to the northern reaches of Glen Canyon, or what little remains of it above the reservoir's tranquil waters, and are devoted almost entirely to servicing a seemingly limitless flotilla of party boats, houseboats, speedboats, fishing boats, jet boats, Jet Skis, all of these "water toys" ebbing and flowing, to-ing and fro-ing, crissing and crossing, zipping up and down, over and back, a whole water world of internal combustion engines burning millions of barrels of oil, gas, freon, and Coors in the heady pursuit of . . . fun. Bruce has our own minor contribution to this mayhem—basically a 20-foot plywood platform built atop two steel pontoons with a 60-horse outboard engine, a steering wheel, and a canvas canopy for shade—hauled down to the boat ramp and set afloat, and after a sizable breakfast we make our way cautiously between 500 acres of moored houseboats out into the main channel.

And here we are, floating 250 feet above "the place no one knew," as the canyon was described nearly forty-five years ago by the Sierra Club's exhibit format book *The Place No One Knew: Glen Canyon on the Colorado* (photos by Eliot Porter, sidebar text by everyone from John Stuart Mill to Albert Einstein). The "no one" in the title ignores the Anasazi, the Paiutes, the Utes, and the Navajo, all of them serial, albeit intermittent, inhabitants of the general region for over 1,000 years, but as far as white folks go, true enough, not a lot of us white folks.

John Wesley Powell knew it, of course, and pictured it thusly in his 1875 account, *The Exploration of the Colorado River and Its Canyons*. "Past these towering monuments, past these mounded billows of orange sandstone, past these oak-set glens, past these fern-decked alcoves, past these mural curves, we glide hour after hour, stopping now and then, as our attention is arrested by some new wonder." And obviously the out-of-work Civil War veterans and marginalized mountain men who accompanied him on his two trips down the river knew, though only Frederick Dellenbaugh wrote anything of significance about it in his book *Romance of the Colorado River*, published in 1902.

Frank M. Brown, president of the Denver, Colorado Canyon and Pacific Railroad, and his chief engineer, Robert Brewster Stanton, had to know it, though God only knows what color lenses they were wearing when they floated down in 1889 dreaming about building a rail line down the entire Colorado River that would hook up with one in Grand Junction and open up the whole of the Southwest. Unfortunately, the river below Lee's Ferry is not the same mild stream one encounters above (with the exception of Cataract Canyon), and the Brown component of the Brown-Stanton party terminated in Marble Canyon when the hapless entrepreneur, having neglected to provide life jackets for himself and his ensemble, capsized somewhere below Soap Creek Rapid and drowned.

And my old man knew it, floated it in 1947 with Norman Nevills and published an essay about the journey in the *Atlantic Monthly* called "San Juan and Glen Canyon." Improving on Powell's descriptive skills, he wrote, "The sheer cliffs of Navajo sandstone, stained in vertical stripes like a roman-striped ribbon and intricately cross-bedded and etched, lift straight out of the great river. This is the same stone, though here pinker in color, that forms the domes and thrones and temples of Zion and the Capitol Reef. It is surely the handsomest of all the rock strata in this country. The pockets and alcoves and glens and caves which irregular erosion has worn in the walls are lined with incredible greenery, redbud and tamarisk and willow and the hanging delicacy of maidenhair around springs and seeps."

That was fifteen years before the canyon's carotid artery was, in effect, cut by the Bureau of Reclamation and it disappeared beneath the stagnant, oil-slicked waters of Lake Powell (or Lake Foul, as its detractors prefer to call it), an impoundment that backed up behind Glen Canyon dam, creating 2,000 miles of recreational paradise shoreline and flooding not only the main gorge, but every side canyon, gully, fissure, and ravine that fed into it.

I notice that a National Park Service boat with a water cannon on board is dogging us across the lake as we make for the fuel docks at Hall's Crossing. We have ten or twelve marine gas tanks stacked in the wells on either side of our engine, three-quarters of them empty, and we obviously need them full if we're going to reach the confluence with the Escalante and then go up it to our prime objective, Cathedral in the Desert. The National Park Service rig suddenly speeds past us for maybe 500 yards, and then the two rangers on board start operating their cannon in a wide circular arc, as if engaged in some kind of fire drill. Each time they swing it in our direction they come closer to hosing us down. "They know this boat," Bruce says. "They know it belongs to the Glen Canyon Institute, and they always give us a hard time because they don't like environmentalists."

"Park rangers don't like environmentalists?"

"Not ones who want to decommission Glen Canyon dam, drain the lake, and return the river corridor to its natural state."

When we tie up at the fuel dock the National Park Service boat pulls in behind us, and the constabulary (for that's what they truly are) step smartly over to confront us. They both have buzz cuts and deep tans, and neither seems to have been born with lips. Tweedledum and Tweedledee, though one is graying and clearly older. They also have nasty-looking side-arms that they wear high on their narrow hips, and they are not in the least bit nicey-nice in their comportment. "Get your PFDs and put them on the deck." Pause. "How many flotation cushions have you got?" Pause. "We need to see your Utah boat registration." Pause. "Turn on your running lights. Blow your horn." And so on. But the institute has been through this before, and all their equipment is in order. Not being able to bust us on a technicality pisses off the park police.

One of them begins to interrogate us individually. "Where you from?"

"Utah."

"New Mexico."

"California."

"Colorado."

The merest whisker of a smile appears. He thinks he's got us now, thinks we're tourists who've hired a nonlicensed operator (Bruce) in a noncommercial boat (the Glen Canyon Institute's) to take us on a private tour. "How much you being charged for this trip?"

"Nothing. We're sharing expenses."

"How you all know each other?"

"Well, this guy was in the Peace Corps with that guy, and that guy is the son of this guy, and this is my daughter . . ."

After another five minutes of pointless harassment, Philip Fradkin becomes terminally annoyed and demands to see their Park Service badge numbers, telling them he's going to report their insolence to the superintendent of Glen Canyon Recreational Area (which, in fact, he does). The youngest uncivil servant seems to consider the possibility of shooting Fradkin on the spot, but after a moment curls his non-lip as if he's just encountered something truly revolting, turns his back, and steps off the boat.

A sun-bleached kid at the gas pump hands me a brochure puffing a variety of nearby recreational possibilities. "For a rejuvenation of the mind," it says, "try a 4-day houseboat getaway at picturesque Lake Powell for only $1,390.00."

Now apparently released from our detention, we back away from the dock and chug off down the river. Ten years before the building of Glen Canyon Dam, in September 1953, David Brower, executive director of the Sierra Club, sent my father a film script he was preparing on Dinosaur National Monument to publicize the Bureau of Reclamation's proposal to build the dam in Echo Park. In the accompanying letter he asked about other areas in the Southwest that might be deserving of protection, specifically in the Four Corners region. Pop replied:

> My own feeling, off the cuff, is that the Wayne Wonderland in southern Utah is a good deal more spectacular and more worthy of preservation as a monument and recreation area, but it isn't at the moment or in any conceivable future time likely to be threatened by major dams. Again, the big threat is mineral claims, especially uranium, and the airports and installations that accompany strikes. *More immediately threatened than either of these regions is the Glen Canyon* (italics mine), which ought properly to be added to the Rainbow Bridge National Monument, but which is, I guess, already doomed by dam-makers. You probably know Glen Canyon; for my money it's better than the Yampa-Green section, and it has the Navajo Mountain, Rainbow Bridge as well as the Natural Bridges monument on its fringes. Really sound planning would reserve the natural bridges, Glen Canyon, and the Rainbow Bridge as a single national park—and you could throw in the Four Corners without straining probability too much.

But Brower was too focused on the threat to the Green and Yampa; he did not, in fact, know Glen Canyon, and he and the other conservation organizations fighting against an impoundment in Dinosaur

National Monument eventually traded off their opposition to a dam site on the Colorado for the salvation of Echo Park. Ten years later, Brower would spend twenty-one days in the place that not enough people knew and would write to Wallace Stegner, "There could be no more unconscionable crime against a scenic wonder of the world than Glen Canyon Dam. You said it long ago, and now I know it—Dinosaur doesn't compare."

It doesn't take many river miles to see why he thought so, though it still requires a good imagination. The lake level, 400 feet deep when it is at "full pool," has dropped 145 feet during the drought years, and the sheer, stripped canyon walls of Wingate sandstone once again soar upward into a narrowing space of cornflower blue sky, massive, magnificent, shaded by overhang, flaming where the sun strikes them around the twists and turns, but it is still just a taste of the grandeur that once was, and maybe will be again if global warming will go on diminishing the winter snowpack enough in the mountainous headwaters of the Colorado, Green, and San Juan Rivers so that this cesspool can't recharge.*

And cesspool it is, a toxic, sediment-laden soup (30,000 truckloads of sediment a day) contaminated

---

* The average flow rate of the Colorado by July is 9500 acre-feet; the Green, 7240 acre-feet; the San Juan 2010 acre-feet. Measurements in July 2007 for the Colorado were 3700 acre-feet, or 38 percent of normal; for the Green, 1490 acre-feet, 20 percent of normal; for the San Juan, 829 acre-feet, 41 percent of normal. Lake Powell was 52.12 percent of full. Not as low as the year of our trip, but still promising.

with selenium, boron, lead, mercury, arsenic, and percolated uranium mill tailings from Moab and Hite. In a "dead pool" storage tank like Lake Powell, none of that gets flushed away. Also hard to ignore is the Saran Wrap, the Styrofoam cups, the plastic bags floating by, the flip-flops, baseball caps, lettuce leaves, remains of a sandwich, tin foil, beer cans, pop bottles, all the crap that blows off or is tossed off people's boats. To be sure, the bathtub ring left by the receding waters is wearing off the cliffs, desert varnish is once again streaking the walls, side canyons are reemerging, along with sandbars sprouting new vegetation, but there is definitely room for improvement before "restoration" becomes a viable concept.

Not to mention that for the past four hours there has been a virtual armada of houseboats passing us in both directions, 40-, 50-, 60-footers, even 75-footers some of these behemoths, with two decks and waterslides and Jet Skis hanging in davits, portly gentlemen in flowered shirts and captain's hats, cocktails, bikinied maidens waving cheerfully at our little tub wallowing in the trough they leave as they motor purposefully by. Hi, there. Isn't this glorious.

They have model names, these hulking craft: *The Odyssey*, *The Excursion*. I read in my rejuvenation of the mind pamphlet that the former sports six staterooms with queen-sized beds, two baths (only two?), full kitchen, central heat and air conditioning activated by a 20-kilowatt generator, hot tub, a lounge with a gas log fireplace and a widescreen, LCD home-theater TV fed by a satellite dish—the whole rig powered by twin 220-horsepower, fuel-injected, stern-drive engines. Of course, one has to drive the damn thing oneself and cook one's own meals and pour one's own drink. But for those who don't wish to be bothered, there's the 96-foot *Canyon Princess* that can be chartered, with full-service bars on two levels and climate-controlled cherry wood salons to sprawl about and relax in.

We reach the Escalante around mid-afternoon and turn up its diminished but still flooded channel, keeping our eyes peeled as we go for a quiet side canyon with a suitable sandbar on which to pull up and make camp. Yeah, right! Us and every other yahoo in Kane County, Utah, most of whom seem to be moored for the duration. Kids squealing down the waterslides, Mom and Sis having one more drink before thawing supper, Pop and Junior back out on the lake chasing one another in madcap circles on their Jet Skis. Generators drone around every bend . . .

So, to make a long story short . . . we camp, we eat, we sleep, and in the morning make our way to Cathedral in the Desert, that deeply scoured amphitheater of curving, arching, salmon-colored walls, so overhung in their convexity and concavity that one can barely see the sky, and most times not even barely; that huge domed cavern with its once-upon-a-time hanging gardens of maidenhair fern, monkey flower, alcove columbine, death camas, and its slot-in-the-rampart waterfall of a manifestly gynecological stamp falling into a plunge pool some 30 feet below. We find geometric compositions where random fracturing of the cliff walls has caused great slabs to peel off, leaving underlayment windows of fresh pink canvas crying out for the petroglyphic imprimatur of one's initials, or Kokopelli's genitalia, I♥Judy, Jesus Saves.

At least that's how I imagine it back when one walked up a half-mile creek bottom from the Escalante into this chancel, back in the days before all these side gulches, Coyote Canyon, Hurricane Wash, Davis Canyon, were flooded by Lake Powell. And in its newly re-revealed, albeit temporary state, Cathedral in the Desert is still deserving of its iconic standing. But we can't quite ignore the fleet of motorboats already tied up at the base of the waterfall (which recently rising waters have foreshortened to less than 10 feet and will soon obliterate), and the number of ovoid hominids climbing up its labial walls, and the racket they make hooting and hollering from the sheer distraction of it all, and the static emanating from the ship-to-shore radio on some boob's Sea Ray 280 Bow Rider. Would they behave this way in Westminster Abbey? I wonder what Major Powell and his men would think about all this.

What *I* begin to think about all this is *get me out of here*. Lake Powell may have suffered a major drawdown and over time global warming may continue the process, siltation may eventually force a decommissioning of the dam as a source of hydroelectric power, and Western water managers may wisely decide that it makes more sense to store water in Lake Mead and keep it full rather than sustain two reservoirs at less than half full. Hundreds of miles of side canyons may in fact enjoy full recovery at some point, which is hardly something to stay crabby about. But for my money, unless Glen Canyon ever again becomes an actual *canyon*, as in a deep fissure in the earth with nothing at the bottom of it but a narrow riparian corridor and 6 feet of silt-laden river, it can never be thought of as restored.

But who knows. Just yesterday they touched off 4,000 pounds of explosives out in Oregon and blew a big hole in the Marmot Dam near Portland, stage one of a deconstruction that when complete will, for the first time in a century, leave the Sandy River free-flowing all the way from Mt. Hood to the Pacific Ocean. Years of talk transformed into action, and it wasn't the Monkey Wrench Gang that did it; it was Portland Power. Maybe one day the same will happen on the Colorado.